Research in Human Development

Volume 1, Number 4

Special Issue: Risk and Resilience in Human Development
Guest Editor: Corey L. M. Keyes

NEW YORK AND HOVE

First published by Lawrence Erlbaum Associates, Inc., Publishers
10 Industrial Avenue
Mahwah, NJ 07430

This edition published 2013 by Psychology Press

Psychology Press	Psychology Press
711 Third Avenue	27 Church Road
New York	Hove
NY 10017	East Sussex, BN3 2FA

Psychology Press is an imprint of the Taylor & Francis Group, an informa business

RESEARCH IN HUMAN DEVELOPMENT, *1*(4), 223–227

Risk and Resilience in Human Development: An Introduction

Corey L. M. Keyes
Emory University

At its most elemental level, human development consists of constancy and change in quantitative and qualitative aspects of behavior and functioning. The science of human development therefore illuminates the causes, mechanisms, and consequences of constancy and change in behavior and functioning. There is no simple characterization of human development other than complex because it is a multidimensional (i.e., biological, social, emotional, and cognitive), multidirectional (i.e., positive and negative), multidetermined (i.e., self, others, contexts, and systems), life-long process (Baltes, Staudinger, & Lindenberger, 1999; Erikson, 1950, 1959; Mortimer & Shanahan, 2003). This special issue of *Research in Human Development* is devoted to the topic of "risk and resilience" in human development, a topic that epitomizes the complexity of human development as a process of constancy and change throughout life.

The terminology of risk and especially resilience is everywhere these days, suggesting these are popular fields of social scientific inquiry that have been ongoing for many years. However, it is important to note that although popular, the topics of risk and resilience are relative newcomers to the field of human development. The terminology and science of risk and resilience emerged primarily from the health sciences and particularly from the investigation of psychopathology. It was a handful of pioneering psychologists—Norman Garmezy (1973), Emmy Werner (Werner & Smith, 1977), and Michael Rutter (1979)—primarily interested in child development who bridged the study of risk and resilience and brought this dual focus into the science of human development.

Risk factors are causes of undesirable, non-normative developmental outcomes. Put differently, risk factors generate negative change in or persistent (i.e., chronic) poor behavior or functioning. Risk factors are measurable characteristics

Requests for reprints should be sent to Corey Keyes, Emory University, Department of Sociology, Room 225 Tarbutton Hall, 1555 Dickey Drive, Atlanta, GA 30322. E-mail: corey.keyes@emory.edu or qualities of individuals, interpersonal relationships, contexts, and institutions. Changes or levels of these characteristics or qualities temporally precede the changes in quantity or quality of a developmental outcome. Hence, a risk factor elevates the probability of the occurrence of a discrete outcome (e.g., disease or illness) or the probability of change in levels of an outcome (see Kraemer et al., 1997; Mrazek & Haggerty, 1994).

Resilience is a risk factor that has been averted or unrealized. The years of research that preceded the study of resilience led to a growing consensus on risk factors for specific psychopathologies (see, e.g., Mrazek & Haggerty, 1994) that include attributes such as poor social skills, exposure to violence or war, low education, and persistent poverty. Individuals who possess, are exposed to, or reside in known risk factors for a sufficient period of time are said to be "at risk" for the undesirable developmental outcomes. Repeatedly, research has demonstrated that at-risk individuals were more likely to develop undesirable developmental outcomes than individuals without or not exposed to the risk factor. However, not all individuals at risk developed psychopathologies or problems. In fact, enough individuals exposed sometimes to substantial or extreme adversities (e.g., physical trauma) developed normatively and even sometimes developed exceptionally. Resilience, according to Masten and Coatsworth (1998), is a pattern of behavior and functioning indicative of positive adaptation in the context of significant risk or adversity. The study of risk and resilience seeks to illuminate the causes, mechanisms, and subsequent developmental consequences of constancy (i.e., normative outcomes or "doing okay") and change (viz. recovery from episode of negative change or growth and improvement) in behavior and functioning in the face of adversity.

Garmezy (1974; Garmezy, Masten, & Tellegen, 1984) and Rutter (1985, 1990) have studied the development of children of mothers with severe mental illness. Research had shown that such children were at risk for undesirable developmental outcomes (i.e., less competence and more disruptive behaviors) when compared with children of mothers without mental illness. Nonetheless, many at-risk children developed normally in terms of developing levels of various kinds of competence (as judged by teachers, peers, and school records) that were similar to the competence of children without the risk factor. Werner (1995) and Werner and Smith (1977) have studied over three decades of the development of a cohort of children born in Kauai (Hawaii), a third of whom were born into household poverty, parental psychopathology, and familial discord. Approximately a third of at-risk children grew into adults who exhibited average or above-average levels of confidence, capacity to care for others, and competence. The looming question is how individuals are resilient, or what makes some resilient and others vulnerable to adversity.

Similar to stress research, which has focused on "buffering effects" of stress, developmental scientists have conceptualized the causes of resilience as "protective factors." Whether it resides at the individual, familial, or community level of analysis, a protective factor moderates the usual statistical relation between a risk factor and an undesirable outcome. Over the past 30 years of investigation, research has identified several key protective factors. At the individual level, resilience has been attributed to high IQ, problem-solving competence, high self-efficacy, and personalities that are autonomous, active, outgoing, and warm. At the familial level, researchers have identified high family cohesion, social support, high-quality parenting, stable family units, and higher socioeconomic standing (e.g., higher household income) as enablers of resilience. Finally, at the level of the community, resilience has been attributed to counseling and support programs and good schools (see Masten & Powell, 2003, for a more complete review of protective factors).

Notwithstanding the growing corpus of findings, the field of risk and resilience research in human development has recently undergone a period of internal, critical reflection. Recent reviews by leadings scholars in this field (Luthar, Cicchetti, & Becker, 2000; Luthar & Zelazo, 2003) have revealed a core set of concerns and potential weaknesses that reflect external and internal criticism of this important field. That is, many of the leading scholars in his field are calling for (a) greater operational consensus, (b) more application and development of theory, (c) explication of mechanisms linking cause and outcome, and (d) extension of research on risk and especially resilience throughout adulthood. Thus, there is concern that too much of the research consists of heterogeneity in the conception and operationalization of resilience, too little theory-driven investigations, not enough research on mediators linking risk to outcome, and an almost exclusive focus on youth. To be fair, such criticisms are not unique to resilience research but have been leveled at almost all domains of scientific inquiry at one point in time. The latter criticism is at once a reflection of the tendency of U.S. research in human development to be age segregated (see Ryff & Singer, 2003) as well as the practical issue that some of the longitudinal studies of resilience will "mature" into adulthood as their cohorts mature as well. With consistent funds to continue these important longitudinal studies, resilience research will truly become the study of resilience as a lifelong process.

The three empirical articles in this special issue represent strong contributions to the growing corpus of research on risk and resilience in human development. Spencer, Fegley, Harpalani, and Seaton focus on the uniquely vulnerable population of urban ethnic minority adolescent males. This population, especially African American males, is at serious risk for premature mortality due to violence or suicide, incarceration, and school dropout or underachievement. This article by Spencer et al. descriptively introduces a potentially new mediator that may link the adversities experienced by urban minority young males with adverse outcomes, which is the construct of hypermasculinity. Most important, Spencer et al. demonstrate that resilient males (i.e., average school grades of an "A" or "B") had lowers levels of hypermasculinity and more positive self-images than the males described as "marginally achieving" (i.e., average school grades of "C" or "D"). Indeed, Spencer et al. also report a positive correlation between hypermasculinity and negative reflected appraisals from teachers and peers, which in turn correlated negatively with the measure of positive self-image.

The final two articles focus on mental disorders, the first viewing it as a developmental outcome and the final article as a risk factor for subsequent developmental outcomes. Feldman, Conger, and Burzette focus on the risk of and resilience from trauma in a sample of mostly White individuals from rural Iowa who were, at the last assessment, in the midst of the transition into young adulthood. Feldman et al.'s article measures an array of mental disorder outcomes and reveals the theoretical importance of developmental timing of the risk factor. That is, Feldman et al.'s article demonstrates that childhood traumas were more strongly predictive of mental disorders than trauma experienced during adolescence. However, a number of young adults were resilient in that they had experienced trauma but were not diagnosed with a mental disorder. Feldman et al.'s article reveals that increased social support from more sources (i.e., family, friends, and school personnel) operated as a protective factor but only for certain mental disorders, another finding that should encourage theoretical development.

The article from Gralinski-Bakker, Hauser, Stott, Billings, and Allen is a study begun in 1978 of a cohort of individuals with a serious (i.e., required hospitalization) adolescent-onset mental disorder. The analyses in this article focus on risk and resilience in this cohort of individuals who are now in the midst of adulthood (i.e., in their "30s") and compared this cohort against the risk and resilience profiles of a matched sample without a serious adolescent-onset psychiatric disorder. Early onset mental disorders are risk factors for adult social role dysfunction (e.g., unemployed, unplanned or early childbirth, and divorced or unmarried) as well as decreased subjective well-being. As such, Gralinski-Bakker et al. conceptualize resilience outcomes both objectively in terms of successful adult social roles and subjectively in terms of psychological well-being. Gralinski-Bakker et al.'s article reveals important gender differences in resilience; women with a prior psychiatric disorder were more likely than the men to exhibit profiles of resilience. Moreover, Gralinski-Bakker et al.'s article suggests possible mechanisms linking this gender difference in resilience outcomes, implicating difficulties such as drug use during the period of emerging adulthood (i.e., the "20s").

In sum, the study of risk and resilience is now a vibrant and significant part of the study of human development. Whether it is the need for operational rigor, theoretical substance, mechanistic linkages, or resilience as a process of adulthood, each article in this special issue adds to the corpus of the frontiers of resilience research. On behalf of the Society for the Study of Human Development (SSHD), I thank the authors whose contributions to this journal support the aspirations of SSHD. Moreover, I am grateful to the outside reviewers for their time and expert recommendations and to Katie Connery at Tufts University for her management of the journal operations and guidance on this special issue.

REFERENCES

Baltes, P. B., Staudinger, U. M., & Lindenberger, U. (1999). Lifespan psychology: Theory and application to intellectual functioning. *Annual Review of Psychology, 50*, 471–507.

Erikson, E. H. (1950). *Childhood and society*. New York: Norton.

Erikson, E. H. (1959). Identity and the life cycle. *Psychological Issues, 1*, 18–164.

Garmezy, N. (1973). Competence and adaptation in adult schizophrenic patients and children at risk. In S. R. Dean (Ed.), *Schizophrenia: The first ten Dean Award lectures* (pp. 163–204). New York: MSS Information.

Garmezy, N. (1974). The study of competence in children at risk for severe psychopathology. In E. J. Anthony & C. Koupernick (Eds.), *The child in his family: Children at psychiatric risk* (pp. 77–97). New York: Wiley.

Garmezy, N., Masten, A. S., & Tellegen, A. (1984). The study of stress and competence in children: A building block for developmental psychopathology. *Child Development, 55*, 97–111.

Kraemer, H. C., Kazdin, A. E., Offord, D. R., Kessler, R. C., Jensen, P. S., & Kupfer, D. J. (1997). Coming to terms with the terms of risk. *Archives of General Psychiatry, 54*, 337–343.

Luthar, S. S., Cicchetti, D., & Becker, B. (2000). The construct of resilience: A critical evaluation and guidelines for future research. *Child Development, 71*, 543–562.

Luthar, S. S., & Zelazo, L. B. (2003). Research on resilience: An integrative review. In S. S. Luthar (Ed.), *Resilience and vulnerability: Adaptation in the context of childhood adversities* (pp. 510–549). New York: Cambridge University Press.

Masten, A. S., & Coatsworth, J. D. (1998). The development of competence in favorable and unfavorable environments: Lessons from research on successful children. *American Psychologist, 53*, 205–220.

Masten, A. S., & Powell, J. L. (2003). A resilience framework for research, policy, and practice. In S. S. Luthar (Ed.), *Resilience and vulnerability: Adaptation in the context of childhood adversities* (pp. 1–25). New York: Cambridge University Press.

Mortimer, J. T., & Shanahan, M. J. (Eds.). (2003). *Handbook of the life course*. New York: Plenum.

Mrazek, P. J., & Haggerty, R. J. (Eds.). (1994). *Reducing risks for mental disorders: Frontiers for prevention intervention research*. Washington, DC: National Academy Press.

Rutter, M. (1979). Protective factors in children's responses to stress and disadvantage. In M. W. Kent & J. E. Rolf (Eds.), *Primary prevention in psychopathology: Social competence in children* (pp. 49–74). Hanover, NH: University Press of New England.

Rutter, M. (1990). Psychosocial resilience and protective mechnisms. In J. E. Rolf, A S. Masten, D. Cicchetti, K. H. Nuechterlein, & S. Weintraub (Eds.), *Risk and protective factors in the development of psychopathology* (pp. 181–214). New York: Cambridge University Press.

Ryff, C. D., & Singer, B. (2003). Flourishing under fire: Resilience as a prototype of challenged thriving. In C. L. M. Keyes & J. Haidt (Eds.), *Flourishing: Positive psychology and the life well lived* (pp. 15–36). Washington, DC: American Psychological Association.

Werner, E. E. (1995). Resilience in development. *Current Directions in Psychological Science, 3*, 81–85.

Werner, E. E., & Smith, R. (1977). *Kauai's children come of age*. Honolulu: University of Hawaii Press.

RESEARCH IN HUMAN DEVELOPMENT, 1(4), 229–257

Understanding Hypermasculinity in Context: A Theory-Driven Analysis of Urban Adolescent Males' Coping Responses

Margaret Beale Spencer, Suzanne Fegley, Vinay Harpalani, and Gregory Seaton
University of Pennsylvania

In this article, we employ Spencer's (1995) Phenomenological variant of ecological systems theory (PVEST) as a framework to examine risk and resilience, with a specific focus on understanding hypermasculine attitudes among low-resource urban, adolescent males. In the article, we highlight the need to understand normative human development processes in context and to consider risk and resilience in conjunction with these processes. We describe findings from a study of risk, social supports, and hypermasculinity. In the discussion, we outline the implications of these findings for theory and practice.

A comprehensive and nuanced understanding of risk and resilience is among the most salient prerequisites for the application of human development research to policy and practice. The risks that youth face, along with the successful and unsuccessful strategies they employ to cope with these risks, need to be understood both in relation to their maturation and identity development and as linked to the social, cultural, and historical context in which these youth develop. Throughout the broad, interdisciplinary realm of human development, it is critical to conceptualize the development of lives in context. It is apparent from research efforts on diverse, urban youth that this is rarely the case as illustrated in the ways that researchers formulate questions, identify constructs, theorize about phenomena, interpret results, and implement social policy. A myriad of conceptual flaws have

Requests for reprints should be sent to Margaret Beale Spencer, GSE Board of Overseers Professor of Education and Psychology, University of Pennsylvania Graduate School of Education, 3700 Walnut Street, Philadelphia, PA 19104–6216. E-mail: marges@gse.upenn.edu characterized such research, particularly with regard to African American youth (Spencer & Harpalani, 2001). Moreover, there are few developmental periods for which this shortsightedness continues to have more dire consequences than adolescence—in the formulation of research, theory, practice, and policy.

Recently, risk and resilience have been popular themes for research in applied human development. Scholars have begun to explore the important conceptual issues that require greater clarity such as the various definitions of resilience as a construct and the utility of these in various contexts (see, e.g., Luthar, 2003; Luthar & Cicchetti, 2000; Luthar, Cicchetti, & Becker, 2000; Spencer, 2001). Our approach to these issues employs an "identity-focused, cultural-ecological" perspective. We examine identity statuses that confer privilege or marginality, such as race and gender, and we consider the impact of these factors along several dimensions: individuals' perceptions of self and future opportunities, imposed expectations or stereotypes that result from inferences about race and gender, and the interaction between both of these factors and normative maturational and development processes such as puberty and identity exploration in adolescence. Risk and resilience cannot be separated from normative human development (which itself involves some level of risk taking and demonstration of resilience); rather, risk is best viewed as an exacerbation of normative challenges and competencies due to larger sociopolitical processes (i.e., racism, sexism) and/or lack of resources and resilience as successful coping with these exacerbated challenges.

Furthermore, we emphasize the need to understand these issues as impacted by multiple levels of context. This includes both the immediate situational influences across various contexts of development, such as school, family, and neighborhood, and the larger societal influences including political decisions and media messages that are filtered through the more proximal venues. The interactions between these niches and levels of context should also be considered when designing interventions and policies to help youth cope with traditional developmental challenges and with exacerbated normative adolescent tasks.

In this article, we present a conceptual and empirical demonstration of how to study urban adolescents' lives in context. First, we present our conceptual framework, Spencer's (1995) phenomenological variant of ecological systems theory (PVEST), which aims to integrate the salient, aforementioned issues of normative development and context. We then discuss one particular coping response, hypermasculinity, as it is impacted by various risks and contextual factors. Subsequently, we apply the PVEST framework to examine how risk factors such as race and socioeconomic status (SES) impact the formation of hypermasculine attitudes among urban, low-resource, adolescent males in conjunction with available social supports across family and school settings. We also consider these factors in relation to positive views of self, an adaptive coping outcome. Given that educational performance is critical for adulthood competence and is also related to risks, social supports, and psychosocial outcomes (Connell, Spencer, & Aber, 1994), it can serve as one partial proxy for resilience. Our analysis includes one relatively resilient (high academic achieving) group and one relatively vulnerable (marginally academic achieving) group.

Although our focus in this article is hypermasculinity, which is often conceived as a maladaptive response (although not always, as we discuss), it is necessary to consider positive and negative coping strategies in relation to each other. Elsewhere, researchers have discussed positive coping strategies employed by urban, African American youth including proactive Afrocentric cultural identity (Spencer, Noll, Stoltzfus, & Harpalani, 2001) and spiritual coping (Spencer, Fegley, & Harpalani, 2003). Here, we also study hypermasculinity in conjunction with more positive entities including social supports and positive views of self. Nevertheless, we do focus on hypermasculinity because it is not well understood and often misrepresented. Our study illustrates the need for a broadened conceptualization of the salient issues, and our discussion includes suggestions for policy and program innovations of particular relevance for understanding urban adolescents' lives in context.

CONCEPTUAL AND EMPIRICAL OVERVIEW

The Phenomenological Variant of Ecological Systems Theory

Spencer's (1995) PVEST combines a phenomenological perspective with Bronfenbrenner's (1989) ecological systems theory, linking culture and context with individuals' meaning-making processes and resultant identity formation. Bronfenbrenner's model provides a very useful conceptualization of the multiple levels of context: microsystems (family, school, etc.), mesosystems (relations between various microsystems), exosystems (external venues such as parent's employment setting), and macrosystems (government policies, broad social and cultural influences, etc.). Essentially, PVEST aims to elucidate how individuals experience and are affected by these multiple levels of context throughout the life course. Rather than describing context directly, as Bronfenbrenner's model does, PVEST illustrates and examines normative human development processes—framed through the interaction of identity, context, and experience. It considers differences in experience, perception, and negotiation of stress and dissonance (or lack thereof). The components of PVEST also allow one to see how normative developmental challenges are exacerbated by risk factors and how resilience is related to the strategies that youth employ to cope with these exacerbated challenges. Thus, PVEST utilizes an identity-focused cultural-ecological perspective, integrating issues of social, political, and cultural context with normative developmental processes.

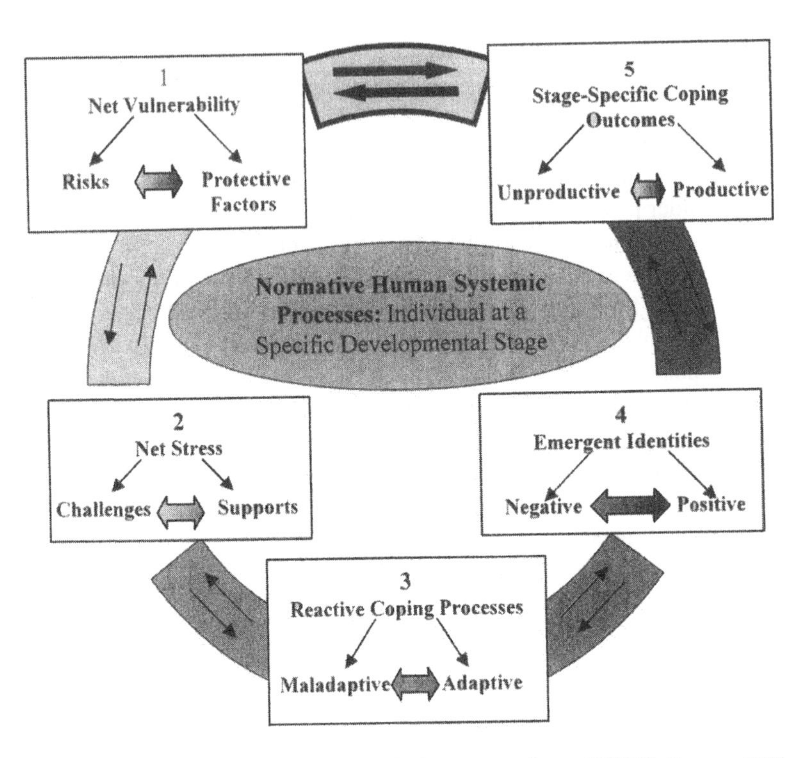

FIGURE 1 Phenomenological variant of ecological systems theory (PVEST; Spencer, 1995, revised in 2004).

PVEST consists of five basic components that form a dynamic, life course model (Figure 1). The model seeks to make explicit the dynamic relations between (1) risk, (2) stress, (3) coping, (4) identity, and (5) life outcomes. The first component, net vulnerability (1), consists of the contexts and characteristics that may potentially pose challenges during an individual's development. Risk contributors are factors that may predispose individuals for adverse outcomes. The risks, of course, may be offset by protective factors, thus defining net vulnerability for a given individual. For many urban youth, neighborhood dangers pose numerous risks to healthy development, and these must be offset by appropriate protective factors to insure positive outcomes.

Net stress engagement (2), the second component of PVEST, refers to the actual experience of situations as perceived by the individual that challenge the individual's well-being; these are risks that are actually encountered and must be dealt with. Available social supports can help youth negotiate experiences of stress; thus, supports are actualized protective factors. Whereas risks and protective factors denote potential entities within the environment, stress and support refer to actual manifestations of these entities—experiences in context. Neighborhood dangers that are actually encountered and experienced (again, as perceived by the individual) constitute stressors. Thus, PVEST forges a link between context and experience.

In response to stressors and in conjunction with supports, youth employ reactive coping methods (3) to resolve dissonance-producing situations. Normative cognitive maturation makes awareness of dissonance unavoidable and acute. Reactive coping responses include problem-solving strategies that can lead to either adaptive or maladaptive solutions.

As youth employ various coping strategies, self-appraisal continues, and those strategies yielding desirable results for the ego are replicated. Accordingly, they become stable coping responses and coupled together yield emergent identities (4). Emergent identities define how individuals view themselves within and between their various contexts of development (e.g., family, school, neighborhood). The combination of cultural/ethnic identity, sex role understanding, and self-appraisal and peer appraisal all define one's identity.

Identity processes are of critical value in that they provide behavioral stability over time and space. Identity lays the foundation for future perception, self-appraisal, and behavior, yielding adverse or productive life-stage-specific coping outcomes (5). Productive outcomes suggest competencies that include good health, high academic performance, positive relationships, and high self-esteem, whereas adverse outcomes include poor health, incarceration, and self-destructive behavior.

The PVEST framework represents dynamic processes that continue throughout the life span as individuals balance new risks against protective factors, encounter new stressors (potentially offset by supports), establish more expansive coping strategies, and redefine how they view themselves, which also impacts how others view them. Unresolved issues within one life stage influence the character and level of stress experienced and also impact future coping and identity formation processes (Erikson, 1968). PVEST aims not only to capture this developmental process but also to place it within the broader social, cultural, and historical contexts.

Masculinity and Adolescent Coping in Context

As noted, human development and maturation are directly linked to context. More specifically, gender identity development (the process of understanding oneself in gendered terms) is linked to context. The economic decline of cities has significantly influenced how urban boys understand themselves in gendered terms and how they fulfill gendered social roles. Since the 1980s, major American cities have experienced tremendous decline due to the closing of many industrial and manufacturing companies. The loss of jobs has created extremely high levels of unemployment and serious decline of the urban infrastructure (Wilson, 1996). As a result, poverty and decline have become synonymous with urban America. Given the rapid decline of urban cities and rise of underground drug economies, many urban residents, young and old alike, are increasingly more concerned about their personal safety (May, 2001). Despite a downward trend in crime statistics, fear of victimization has remained constant and in some cases inordinately high. Increased fear of victimization is in large part due to the better and increased modes of mass communication as well as the continued racialization of crime and criminality. For example, Glassner (1999) reported that over the last 5 years, violent crime decreased by 6%; however, media coverage of violent crime increased by 300%. Although reported crime is on the decrease, neighborhood and school quality either remains stagnant or becomes progressively worse. Actual violence (or even the perceived threat of violence) is a stressor that presents serious psychological and physical risks to adolescents. To develop appropriate intervention strategies, a clear understanding of adolescent identity development, fear of crime, and cultural context are required.

Masculinity in America is already a problematic proposition, as it robs many boys of their emotional health. Masculine norms tend to discourage the display of vulnerability; consequently, many adolescent males adopt a presentation of self that may seem confident and stable when in fact, internally, this may not be the case. In an exaggerated form, this discordance of emotions may manifest itself as *hypermasculinity*: the exhibition of stereotypic gendered displays of power and consequent suppression of signs of vulnerability. This exaggerated presentation of masculinity can lead to conflict in school, neighborhood, and family settings, but it can also serve as a coping response to deal with environmental stressors such as lack of economic opportunities and fear of victimization. Male identity development is influenced by ecology, and many urban adolescent males must meet the challenges of high-risk contexts in addition to normative, societal expectations of masculinity. Whether it is actual physical spaces (neighborhoods or schools) or social structures (i.e., unemployment or racism or culture), ecology matters in the lives of developing youth, and the impact of masculine norms, which as noted may be problematic in and of themselves, is exacerbated by ecological risks—contingent on available social supports.

Human interaction is central to each sphere or layer of context, and cultural factors permeate each level. Cooley (1902) posited that individuals see themselves and others as symbols in their social environment—the notion of symbolic interaction. Context is saturated with symbolic representation. In considering neighborhoods and schools, physical appearance as well as race conveys a symbolic message. For example, the "broken window" theory suggests that visible signs of neighborhood decay serve as symbolic authorization for criminal, antisocial, and aggressive behaviors (Sampson & Raudenbush, 1999; Shaw & McKay, 1946; Snell, 2001). Consequently, individuals perceive that behaviors such as loitering, littering, and drug dealing are condoned or at least tolerated in some schools or neighborhoods more than in others. Perception and meaning making are central to this process, as ecological context is evaluated not only by potential victimizers but also by all individuals as they assess their respective risks (Ferraro, 1995). Nonetheless, many studies of social disorganization fail to include variables that address the phenomenological perspectives of adolescents.

With regard to hypermasculinity, Mosher and Sirkin (1984) suggested that macho personality was identifiable along three characteristics: (a) view of violence as manly, (b) perception of danger as exciting, and (c) callousness toward women. Traditionally, exaggerated masculinity has largely been viewed as a countercultural phenomenon endemic to Black and Latino males. This "Black and Brown" version of manhood has interchangeably been described in the literature as machismo, bravado, macho personality, and hypermasculinity. Also, some have suggested that hypermasculine behavior is a function of race and culture (e.g., D'Souza, 1995; Moynihan, 1965).

Few studies have attempted to demonstrate the universality of hypermasculinity across race and class. Such an empirical omission perpetuates stereotypes about Black and Latino males by stigmatizing their aggressive behavior as problematic while allowing aggressive behavior among White males to go virtually unacknowledged and unexamined. Consequently, minority males are further criminalized, whereas White males do not receive much needed intervention, and the role of a significant trigger for hypermasculine behavior, fear of victimization, is largely unacknowledged and unaddressed.

Few researchers have examined how adolescent males cope with neighborhood stress specifically as it relates to fear of assault or victimization (May, 2001). The existing literature has not adequately addressed the autonomy of adolescents that is often manifested as increased time away from home. Adolescents travel independently to and from school as well as to extracurricular and leisure activities. Moreover, unlike adults, most urban adolescents do not have the option of driving, which serves to lessen threatening encounters. They must traverse crowds, trash-littered streets, and abandoned properties by more vulnerable modes of travel such as walking or public transportation.

Further, most studies have been singular in contextual focus—neighborhood or school. Seldom have studies examined both contexts simultaneously with regard to fear of victimization (Welsh, Stokes, & Greene, 2000). Consequently, there is limited knowledge on how fear functions and how identities are constructed across multiple, potentially high-risk contexts. School represents the setting where many youth spend the most time away from home. By omitting school and neighborhood contextual connections from analysis, the impact of contextual stressors may be underestimated. As a result, researchers do not optimally understand the role of environment in conjunction with socialized norms of masculinity as a source of stress for urban adolescent boys.

Masculinity and Hypermasculinity: Race and SES as Exacerbating Risk Factors

Given the charged, negative stereotypes associated with African American males (Ferguson, 2001), hypermasculine behavior poses a particularly salient risk for this group. Black males are imprisoned at six times the rate of White males (Donziger, 1996). This trend is also similar for adolescents. Regarding 1980 drug arrests, Donziger (1996) reported, "arrest for African American youth surged, almost quintupled—to 1,415 per 100,000—while arrest for White youth remained stable . . . this occurred even though White juveniles were using drugs at the same rate" (p. 122). The statistics suggest that factors beyond the actual offenses (i.e., perhaps racial/cultural stereotypes related to hypermasculinity) are responsible for such inordinately high incarceration rates among young Black males. In an attempt to better understand the high arrest rates of urban Black neighborhoods, Sampson (1986) used self-report crime measures, census income data, and police records to determine if juvenile arrest was mediated by neighborhood. This study found that independent of self-reported delinquency, low-poverty areas tended to be more policed than relatively affluent neighborhoods and that police officers tended to be less lenient in poorer neighborhoods. Behaviors that may be viewed as adolescent or "juvenile" in some neighborhoods are perceived as delinquent or pathos promoting in others. Sampson (1986) noted, "the influence of SES on police contact is contextual in nature and stems from an ecological bias with regard to police control as opposed to a single individual bias against the poor" (p. 884). These factors converge to place poor urban youth in the maze of the juvenile justice system. Sampson's (1986) results suggest that the impact of neighborhood on delinquency is potentially overestimated because the roles of officer and policing bias are not included. Further, race was the most important factor in determining whether a juvenile would be charged with a crime or released (see Donziger, 1996). Minority youth were more likely than White youth to face charges for the same offense.

In addition to the normative maturational challenges, Black boys must also deal with prejudice and the negative, stereotypic connotations associated with Black masculinity, which is often inherently viewed as hypermasculinity (Spencer, 1999; Stevenson, 1997). These stereotypes are pervasive, as they cut across multiple ecological settings (i.e., self, family, neighborhood, school, employers, law enforcement, and culture in general).

Hypermasculinity as a Coping Response to Fear

Gender identity development is as much a social process as it is a psychological one. Consequently, context plays a large role not only in shaping one's identity through mediating stress and support but also by informing the way future situations are perceived. Urban schools and neighborhood contexts are similar when it comes to the threat of violence (perceived or actual). Community norms or modes of coping are often translated into school settings such that hypermasculine or aggressive behaviors used in the neighborhood are often transported to schools (Stevenson, 1997).

Due to their respective disorder and interconnectedness, violence against youth in school and in neighborhoods is cause for concern. According to the U.S. Department of Education's report, *Indicators of School Crime and Safety, 2000* (Kaufman et al., 2000), there were a total of 60 violent deaths in U.S. schools in 1997–1998. This number is relatively low compared to the 2,752 nationwide juvenile victims of homicide during the same time frame. However, nonfatal acts of violence within schools were approximately half the number of incidents of nonfatal acts of violence within neighborhoods, 253,000 and 550,000, respectively. Although there is a greater probability for violent victimization outside of school, the data suggest that the threat of violent victimization is consistent across neighborhood and school contexts. However, the threat may vary across these contexts as a function of adolescent development; Kaufman et al. (2000) noted, "younger students (ages 12 through 14) were more likely than older students (ages 15 through 18) to be victims of crime at school . . . older students were more likely than younger students to be victimized away from school" (p. viii).

The statistics suggest that both schools and neighborhood can be extremely stressful contexts for adolescents. They must concern themselves not only with normative stressors (i.e., dating and career exploration) but also with experiences of fear with regard to their own person safety. Given such high rates of victimization across both school and neighborhood contexts, fear of victimization in their own neighborhoods and schools may be a salient stressor for many youth, particularly Black males.

Fear is one of the most basic emotional states; it is evoked when one perceives a threat whether real or imagined, and it constitutes an adaptive survival response by prompting the individual to avoid potential dangers. In response to chronic sources of fear, individuals may cope by adopting psychological postures that diminish the possibility of being victimized. For males, these coping strategies include maladaptive aggressive and hypermasculine behaviors. If masculinity is predicated on boys' not acknowledging fear, then boys may not openly express fear, instead attempting to hide and deny it. Consequently, one confounding factor in research studies may be lack of variation on fear measures. Although a legitimate concern, an international survey of youth fear of crime (see Goodey, 1997) reported that 42% of the boys reported being fearful, suggesting that boys will report fearfulness on some measures (perhaps when protected by anonymity of the survey).

With regard to urban youth, the current literature on fear has two serious shortcomings. First, although responses to fear are varied, the literature has, by virtue of its omission of males and youth, presented a monolithic fear response. Almost exclusively, research on fear as a stressor has focused on female and elderly populations. Certain adolescents, often Black males, are viewed solely as the source of fear for adult females and the elderly. Second, too few researchers have examined which environmental characteristics create fear among adolescent males, and researchers have also neglected to study how males cope with fear. As noted, we posit that expressions of hypermasculinity may be one such coping method.

There is a considerable literature with regard to fear of victimization and neighborhood disorder. The concept of neighborhood fear depends primarily on perceptual, interpretive, and meaning-making processes (Ferraro, 1995). Thus, phenomenology is fundamental to any investigation of fear, as individuals must perceive, interpret, and create meaning out of their respective social spaces and its inhabitants. Considerable attention has been given to the issues of research validity regarding fear. Researchers have critiqued the literature and have suggested that a greater distinction between perceived risk and actual risk of victimization is necessary (e.g., Snell, 2001). In many cases, research participants tended to be much more afraid of crime than they were statistically probable of experiencing. For example, during the Washington, DC metropolitan area sniper shootings in 2002, far more people feared being shot than realistically possible for the beltway snipers to actually shoot. Nonetheless, the thoughts and emotions surrounding perceived danger are often as salient as the danger itself. The phenomenological experience of fear, whether danger is real or perceived, must also be considered in conjunction with normative adolescent development—which involves experimentation with gender-specific roles, behaviors, and expressions. Also, for many Black males residing in poor urban neighborhoods, fear of victimization is not only "real" in a phenomenological sense; it is also real in a probabilistic sense as well. Black males, particularly adolescent Black males, have victimization rates higher than any other racial or age group (Donziger, 1996).

Neighborhood effects in general and disorganization specifically are central to any discussion of youth identity development, as these directly influence youth stress and supports. Neighborhood and school have serious implications for the day-to-day experiences of youth, and encounters with stress play an integral role in identity development and life outcomes. For example, Cunningham (1999) found that Black males were very aware of environmental dangers both proximal and distal, and cognitive appraisals of risk were significantly strong predictors of hypermasculine coping (i.e., violence as manly). Cunningham's (1999) study suggests that hypermasculine behavior is a coping response to perceived neighborhood threat and danger.

Stevenson (1997) described how many Black youth reside in high-risk neighborhood contexts where anger display may be an adaptive coping mechanism to deal with fear; indeed, anger may be a form of competence for social and emotional viability. However, the same anger display may constitute hypermasculine behavior with all of its risks. Thus, the negotiation of fear in high-risk neighborhood contexts through the adoption of hypermasculinity as an identity reinforce negative stereotypes and possibly lead to negative outcomes such as adjudication and incarceration. Also, hypermasculine posturing necessary to avoid victimization may also mitigate school adjustment (Spencer, 1999)—by means such as reinforcing negative teacher perceptions. Moreover, these kinds of phenomena must be understood in light of the salient, existing negative stereotypes of Black males, which are already rooted in hypermasculine imagery (Ferguson, 2001).

May (2001) observed similar concerns about personal safety when he examined adolescent (13–19 years old) male weapon possession by level of neighborhood disorder. In a study of 318 males (48% White, 41% Black, and 11% other), May found significant connections between neighborhood disorder, fear, and defensive behaviors that are typically conceived of as hypermasculine (i.e., gun possession, gang participation, and weapon possession). Regarding age, younger males who came from families receiving public assistance were more fearful than older males. With respect to race, there was a significant relation between fear and defensive weapon possession. Black males were significantly more fearful than their White counterparts and resorted to carrying more lethal defensive weapons: guns rather than knives, razors, and sticks. May's (2001) results suggest an alternative explanation to male delinquency in general and Black hypermasculinity more specifically. Perhaps these are not merely attributes of gang membership or urban, Black culture but rather responses to youths' perceptions of neighborhood context and fear.

As noted, coping responses can be either adaptive or maladaptive, and this classification is relative to time and space. Maladaptive coping responses (e.g., hypermasculine displays) may provide short-term relief from a particular stressor (e.g., perceived threats). However, in the long run, these maladaptive coping strategies may lead to increased encounters with stressors and consequently increase the likelihood of negative life outcomes. Thoits (1995), in a review of the study of stress, concluded that one's sense of power over a situation or ability to achieve a role determines the extent to which they will utilize "problem-focused coping." Problem-focused coping, a type of adaptive coping, represents a more rational approach to dealing with stress. In the case of youth, adolescents in general and Black adolescents in particular have relatively little "control." For adolescents in general, adults organize much of their days, lives, and opportunities to explore roles. Further, low-income urban youth may have relatively little control over the neighborhood and school dangers they encounter. Moreover, Black youth have no control over the vast media expressions and stereotypes of Black hypermasculinity, which they interpret in relation to self. Consequently, when manhood is threatened, youth encountering these varying levels of risk may resort to more impulse-driven hypermasculine behaviors.

Spencer has argued that repeated coping strategies used over time become more stable as "emergent identities," noting that "when deployed consistently, the reactive coping becomes firmly internalized with stability across context as emergent identities" (Spencer, 2000, p. 108). Spencer (1999) explored the issue of school adjustment as a potential source of stress engagement for African American youth. *School adjustment* was defined as "the degree of school acculturation required or adaptations necessitated for maximizing the educational fit between the student's qualities and the multidimensional character and requirements of learning environments" (Spencer, 1999, p. 43). This is an area that has not been explored extensively with regard to minority youth.

Focusing on African American males, Spencer (1999) explored the impact of several contextual stressors, including perceived peer unpopularity, perceived negative teacher perceptions, and parental monitoring, on the development of hypermasculine identity. Coping methods, such as aggressive attitude and learning attitude, were also incorporated in this analysis. Spencer (1999) found that parental monitoring was significantly and inversely related to hypermasculinity. The findings also suggested a trend-level direct relation among inferred negative teacher perceptions, perceived unpopularity with peers, and hypermasculinity. Additionally, aggressive attitude was also related to hypermasculinity. These factors can affect both high- and low-achieving Black youth. Spencer (1999) noted that the link between negative teacher perceptions and hypermasculinity suggests the need for mentoring initiatives and programs that facilitate productive partnerships between youth and adults. These partnerships can be very salient in promoting positive outcomes.

Hypermasculinity is not a normative, cultural behavior—as it is often represented in popular media portrayals of Black men. Rather, it is reflective of a coping response to particular contextual stressors. For male adolescents in high-risk contexts, hypermasculinity may be a way of symbolizing gender identity achievement. At the same time, hypermasculinity functions as an individual response to perceived threats or challenges. May's (2001) study supports Spencer's (1995, 1999) emergent identity premise; youth who cope with fear by carrying a weapon or joining a gang often find themselves in situations where they in fact have to use the weapons or engage in deviant behaviors. Sometimes weapons provide a false sense of security, or youth engage in delinquent behaviors to maintain approval among peers. Ironically, youth who initially carry weapons or join gangs for protection often find themselves in the actual situations they attempted to avoid.

In sum, the example of hypermasculinity, with all of its complex nuances, highlights the need to view risk and resilience in context and to integrate normative human development processes in the equation. Hypermasculinity is a response to particular risk factors, and in some forms, it may even be an adaptive (resilient) response in the immediate situations at hand. However, it also creates other risks for adolescent males and may compromise resilience in the long term.

Understanding the complex, nuanced nature of coping responses such as hypermasculinity is a key step in designing targeted interventions and policies to mitigate risk and promote resilience. This understanding must include not only an assessment of risk but also an appraisal of social supports and a precise analysis of how these can contribute to resilience. The impact of social supports on hypermasculinity is even less well understood than the effect of risk; it has hardly been addressed at all. It is in this vein that we present this study, which analyzes how risks, social supports, and coping outcomes (attitudes about self) influence hypermasculinity among relatively resilient (high academic achieving) and relatively vulnerable (marginally academic achieving) low-income, urban youth.

THIS STUDY

Project Overview

Participants in this study represent a subsample derived from a larger study of adolescents participating in a longitudinal randomized field trial (Noll, Cassidy, & Spencer, in press). The larger study was designed to evaluate the effect of monetary stipends on academic resilience and adaptive coping among high academic achieving and marginally academic achieving urban youth from low-income families. We began recruiting our first cohort of 9th-, 10th-, and 11th-grade students during the spring of 1999 from 41 public high schools located in a large Northeastern city of the United States. Given our interest in resilience-enhancing influences in ethnically diverse urban adolescents, we established financial and academic criteria for students' participation in the study. We reviewed applications for financial and academic eligibility; high-achieving (A/B grade average) and marginally achieving (C/D grade average) students whose household income met the financial criteria guidelines for the Federal Free Lunch Program (130% of the poverty line) were eligible to participate in the program. The first wave of the study began in the fall of 1999. The final sample of 882 Cohort 1 participants was 68% female, 61% high-achieving, 60% Black/African American, 16% Asian/Asian American, 12% Hispanic/Latino, 8% White/Euro-American, and 4% Other (as indicated by students' self-reported ethnicity/race). When students entered the study, 32% were in 9th grade, 35% were in 10th grade, and 33% were in 11th grade.

Data Analysis Plan

It is important to acknowledge our assumption that the youth in our sample have all been exposed to some level of risk inherent to living in urban neighborhoods, especially considering that they are a group of predominantly young ethnic/racial minority males who all come from low-income families or households. Given that this is an understudied group of youth (especially our group of high-achieving minority males), it is even more important to understand the impact of adolescents' self-reported experiences and coping styles on their ability to successfully navigate the contexts of their daily lives. Specifically, we examined the relation between two components of PVEST, stress engagement and coping, among our sample of adolescent males. Measures of stress engagement were inferred negative perceptions of teachers, inferred negative perceptions of peers, perceptions of neighborhood risk, perceived parental monitoring, and perceptions of support from school personnel and support from family and friends. Measures of coping included exaggerated sex-role presentations (i.e., hypermasculinity) and positive attitudes about self; both of these can also be conceptualized as stage-specific coping outcomes that contribute to long-term life outcomes consistent with the recursive, cyclically reinforcing nature of the PVEST model. We used zero-order correlations among measures of net stress engagement and coping to examine the strength and direction between variables.

Hypotheses

As noted, the variables in our study reflect components of the PVEST framework highlighting the linkages between risk, stress, support, and coping, all as related to resilience. Researchers, educators, and policymakers generally agree that academic success and positive feelings toward self are two key indexes of adolescent resilience. In general, when attempting to deal with the normative (and nonnormative) stresses associated with adolescence, more academically successful students utilize more adaptive coping strategies than do less academically successful students. Thus, we expected that our sample of high-achieving (relatively resilient) males would report more positive attitudes toward self than our sample of marginally achieving (relatively vulnerable) males and that our sample of marginally achieving males would be more likely than their more academically successful peers to employ less adaptive coping strategies as exemplified by their greater endorsement of hypermasculine sex-role attitudes.

Because our study was designed to examine resilience-enhancing influences (social supports) on groups of high-achieving and marginally achieving youth, we expected that there would be significant mean score differences between our samples of high-performing and marginally performing adolescent males on measures of stress and coping. Specifically, we expected to find differences between high- and marginally achieving students' scores on measures that assessed school-related perceptions (i.e., negative teacher perception and perceived support from teachers, principals, and counselors), with high-achieving males perceiving greater support from school personnel and reporting less negatively biased perceptions of teachers' opinions of their personal attributes than their peers who are not experiencing as much academic success. Further, given research documenting a link between parental monitoring and academic achievement (Spencer, Dupree, & Swanson, 1996), we also expected that our samples' high-achieving males would perceive their parents as being more knowledgeable about their activities and whereabouts than would our marginally achieving students.

In addition to our expectations regarding differences between our samples' high-performing and marginally performing adolescent males on measures of stress and coping, we were also interested in exploring links between net stress engagement and coping. Among both high- and marginally achieving males, we expected that higher levels of social support (e.g., receiving emotional support and care from school personnel, family, and friends; perceiving peers and teachers as not having negative opinions and attitudes toward oneself; and feeling that your parents or guardians are knowledgeable about your whereabouts and activities) and lower levels of perceived neighborhood risk (i.e., feeling less anxious about neighborhood threats to one's personal safety) would be related to more positive self-attitudes and less stereotypical sex-role attitudes.

METHOD

Study Sample

The sample for this analysis is comprised of the 239 male participants from the first study cohort (Noll, Cassidy, & Spencer, in press) who completed their baseline assessment. The study sample was 65% high-achieving, 54% Black/African American, 16% Asian/Asian American, 12.5% Hispanic/Latino, 7.5% White/Euro-American, and 10% Other (as indicated by students' self-reported ethnicity/race). At the time of survey completion, 28% were in 9th grade, 39% were in 10th grade, and 33% were in 11th grade.

Procedures

A trained team of research assistants administered annual surveys to students. Survey administrations either took place in large-group settings at geographically central locations, or they took place in smaller settings at students' schools and at our research center. We offered participants $25 incentives to complete their annual surveys and informed them that tokens would be available for students who took public transportation to and from the survey site.

Demographic Variables

We gathered demographic information from two sources, participants' applications and annual surveys. As part of the application process, we asked parents or guardians to provide information about the age, occupation, and yearly income of all individuals living in their household. As noted, all participants were eligible for the Federal Free Lunch program, and 34% of the participants' household incomes were greater than 50% below the Federal Free Lunch guidelines. We also included questions on the annual survey about respondents' age, family household structure, and maternal employment status and education. At the time of baseline survey completion, male respondents' mean age was 15.3 years (*SD* = 1.2); 37% lived in two-parent households, 45% lived in single-parent households (1% father only), and 18% lived with their grandparents or other relatives and adults; 33% of their mothers were not employed, 24% had not completed high school, 21% had completed some post–high school education, and 20% were college graduates.

Measures of Net Stress Engagement and Supports

For each scale or subscale, students' raw scale scores were computed as the unit-weighted sum of salient items and then transformed to area *T* conversion scores, a method of standardizing scores that does not assume linearity. Scores are converted to *T* scores (range 1–99) based on the cumulative frequency proportions of students' raw scores. The Pearson *r* correlation between raw scores and area *T* conversion scores is typically greater than .95.

Inferred negative perceptions of teachers. We assessed students' tendency to make negatively biased inferences about attitudes and beliefs that the typical teacher held toward the respondent using the 12-item Negative Teacher Perception subscale of the Abbott Adjective Checklist (Abbott, 1981). Students rated for themselves on a 3-point scale ranging from 1 ("never") to 3 ("all the time") the descriptiveness of statements such as "My teacher thinks I'm lazy," or "My teacher thinks I do not want to learn." The alpha reliability for this measure was .81, and the raw mean scale score was 1.48 (*SD* = .36). The mean scale score after conversion was 47.46 (*SD* = 15.13).

Inferred negative perceptions of peers. We evaluated adolescents' tendency to make negatively biased inferences about peers' perceptions of the respondent using a 15-item inventory in which respondents rated for themselves, on a 5-point scale ranging from 1 to 5 (1 = "no one," 2 = "a few," 3 = "about half," 4 = "most," 5 = "all"), how many of their peers would agree with statements such as

"You are a person who has very few friends," or "You are a person who gets mad when you don't get your way." The alpha reliability for this measure was .84, and the raw mean scale score was 1.59 (*SD* = .49). The mean scale score after conversion was 48.45 (*SD* = 13.28).

Perceptions of neighborhood risk. We assessed students' perceptions of neighborhood risk using an 8-item version of the Fear of Calamity Scale (Riechard & McGarrity, 1994). Students rated, on a 5-point scale ranging from 1 to 5 (1 = "not at all," 2 = "a little," 3 = "somewhat," 4 = "a lot," 5 = "all the time"), how much they worried about various risks, such as dying young, getting stabbed, or getting beat up, in the neighborhood where they live. The alpha reliability for this measure was .91, and the raw mean scale score was 2.21 (*SD* = .98). The mean scale score after conversion was 44.47 (*SD* = 15.96).

Perceptions of social support. We assessed adolescents' perceptions of school support and support from family, friends, and others using a 17-item Social Support Questionnaire (Sarason, Levine, Basham, & Sarason, 1983). Respondents rated, on a 6-point scale ranging from 1 to 6, how helpful each of 17 individuals, such as mother, father, and best friend, was in providing them with caring and emotional support when they needed it. On the scale, 1–2 represents "do not have or rarely see," 2–3 represents "not helpful at all," 4–5 represents "somewhat helpful," and 5–6 represents "very helpful." The school support subscale contained three items—school counselors, principals, and teachers—and the support from family, friends, and others subscale contained 12 items that included support from various family members, friends, classmates, best friend, and so forth. The alpha reliability for the school support subscale was .80, and the raw mean score was 3.89 (*SD* = 1.31). The mean scale score after conversion was 47.44 (*SD* = 11.25). The alpha reliability for the family and friends support subscale was .75, and the raw mean score was 3.67 (*SD* = .87). The mean scale score after conversion was 48.47 (*SD* = 11.02).

Student perceived parental monitoring. We assessed adolescents' perceptions of how much their parents or guardians try to monitor their activities and whereabouts using a 10-item self-report inventory (McDermott & Spencer, 1995b). Students rated, on a 4-point scale ranging from 1 to 4, how much their parents or guardians try to monitor various aspects of their lives such as "Where you go most afternoons after school," and "What is happening with you at school." On the scale, 1 represents "not at all," 2 represents "a little," 3 represents "tries somewhat," and 4 represents "tries very much." The alpha reliability was .91, and the raw mean scale score was 2.61 (*SD* = .71). The mean scale score after conversion was 46.72 (*SD* = 10.35).

Measures of Coping

Exaggerated sex-role presentations. We used the 30-item, forced choice Hypermasculinity Inventory (Mosher & Sirkin, 1984) to assess respondents' tendency to sanction exaggerated, stereotypical sex-role attitudes. The revised version of the scale (see Cunningham, 1994) was used with middle school and high school adolescent boys. Each item is comprised of two statements: one that represents a stereotypical sex-role attitude and one that does not. Respondents were asked to select the statement that they agreed with most. Selections of statements endorsing stereotypical sex-role attitudes were coded as 1 and those reflecting nonstereotypical sex-role attitudes were coded as 0. Examples of items include "I believe that it's natural for guys to get in fights," or "I believe that physical violence never solves an issue," and "When considering relationships, some females are good for only one thing," or "When considering relationships, all females deserve the same respect as your own mother." The alpha reliability for this measure was .91, and the raw mean scale score was .29 (SD = .16). The mean scale score after conversion was 47.43 (SD = 10.41).

Positive attitudes about self. Positive attitudes about self were assessed using the Hare/Funder/Block Ego-Esteem/Resilience Scale (Block, 1985; Hare, 1977; Hare & Castenell, 1985; Shoemaker, 1980), a self-report inventory comprised of 23 items drawn from the Hare Self-Esteem Scale (Hare, 1977) (items are expressions of individuals' perceptions of others' positive views of self) and the Funder/Block ego-resiliency q-sort items (McDermott & Spencer, 1995a) (items are self-assertions of attitudes and behaviors). Respondents rated for themselves, on a 4-point scale ranging from 1 to 4, the extent to which they agreed or disagreed with statements such as "I am generous with my friends," or "Most of the people I meet are likeable." On the scale, 1 represents "strongly disagree," 2 represents "disagree," 3 represents "agree," and 4 represents "strongly agree." The alpha reliability for this measure was .85, and the raw mean scale score was 3.10 (SD = .34). The mean scale score after conversion was 48.05 (SD = 10.22).

RESULTS

Zero Order Correlations Among Measures of Net Stress Engagement and Coping

High scores on negative teacher perceptions were related to high scores on negative peer perceptions ($r = .33, p < .0001$) and hypermasculinity ($r = .38, p < .0001$) and low scores on perceived school support ($r = -.20, p < .01$) and positive attitudes about self ($r = -.28, p < .0001$). There was no relation between adolescents' scores on negative teacher perception and their scores on perceptions of neighborhood risk, perceived support from family and friends, and perceived parental monitoring. Similarly, high scores on negative peer perceptions were associated with high scores on hypermasculinity ($r = .39, p < .0001$) and low scores on perceived school support ($r = -.16, p < .0001$) and positive attitudes about self ($r = -.27, p < .0001$). In addition, adolescents who scored higher on negative peer perceptions reported significantly higher levels of perceived neighborhood risk than adolescents who reported being viewed less negatively by their peers ($r = .20, p < .01$). There was no relation between adolescents' scores on negative peer perceptions and their scores on perceptions of support from family and friends or perceived parental monitoring. Further, adolescents who scored higher on perceived parental monitoring also scored higher on perceived support from family and friends ($r = .15, p < .05$) and positive attitudes about self ($r = .24, p < .001$). Finally, there was a slight but nonstatistically significant tendency for boys who reported higher levels of perceived neighborhood risk to also score higher on the hypermasculinity inventory ($r = .11, p = .10$) and lower on the measure of positive attitudes about self ($r = -.11, p = .08$).

Demographic Differences Between Groups of High and Marginally Academic Achieving Males

Chi-square analyses were conducted to examine whether there were differences between high ($n = 155$) and marginally ($n = 84$) achieving groups of students by race/ethnicity, maternal education, family structure, and household income. There were no differences in levels of maternal education, family structure, and household income between high and marginally achieving groups of youth. However, there was a slight difference in students' reports of maternal employment status, $\chi^2(1, N = 231) = 4.19, p < .05$. Although the majority of respondents indicated that their mothers were employed (67%), mothers of marginally achieving boys were slightly more likely to be employed than mothers of high-achieving boys (75% vs. 62%). However, the two groups differed significantly by race/ethnicity, $\chi^2(4, N = 239) = 31.25, p < .0001$. Racial and ethnic group percentages for the high-achieving (A/B) participants were 43% Black/African American, 23% Asian/Asian American, 16% Hispanic/Latino, 10% White/Euro-American, and 8% Other. Racial and ethnic group percentages for the marginally achieving (C/D) participants were 74% Black/African American, 2% Asian/Asian American, 7% Hispanic/Latino, 4% White/Euro-American, and 13% Other.

Academic Group Differences on Measures of Stress Engagement and Coping

T tests were conducted to examine mean score differences on measures of stress engagement and coping for A/B and C/D students. As expected, more academically successful (i.e., A/B) students ($M = 49.78, SD = 9.98$) perceived school counselors, teachers, and principals as being significantly more helpful in providing them with emotional care and support than did less academically successful (i.e., C/D) students ($M = 43.04$, $SD = 12.19$), $t(140) = -4.30$, $p < .0001$. Not surprisingly, C/D students ($M = 53.86$, $SD = 13.69$) reported perceiving significantly higher levels of negative teacher perception than did A/B students ($M = 44.00$, $SD = 14.78$), $t(237) = 5.05$, $p < .0001$. There were no significant differences between A/B and C/D students' scores on perceptions of parental monitoring, support from family and friends, and fear of neighborhood risk. When examining academic group differences in coping, marginally achieving males, as expected, were significantly more likely to endorse stereotypical sex-role attitudes ($M = 50.56$, $SD = 8.94$) than their high-achieving male peers ($M = 45.56$, $SD = 10.80$), $t(223) = 3.57$, $p < .001$. Not surprisingly, A/B students reported higher levels of positive attitudes toward self ($M = 49.60$, $SD = 10.23$) than did the C/D students ($M = 45.17$, $SD = 9.61$), $t(236) = -3.25$, $p < .01$.

Regression Analyses Exploring the Links Between Adolescent Males' Levels of Risk, Net Stress Engagement, and Hypermasculine Attitudes

Because of the significant differences between group characteristics and scores on measures of stress engagement and coping among our high-achieving and low-achieving male participants, we conducted two sets of multiple regression analyses (one for high-achieving students and one for marginally achieving students) to examine the links between risk (i.e., selected demographic characteristics), net stress engagement (i.e., negative teacher and peer perceptions, neighborhood risk, emotional support from school personnel and family, friends, and others), and coping (i.e., the tendency to endorse exaggerated or hypermasculine sex-role attitudes and having positive attitudes about the self).

Table 1 displays the findings for our sample of high-achieving male participants in regression analyses examining the relation among risk, net stress engagement, and adolescent males' reactive coping (i.e., hypermasculinity). Age, household income, being Black, and being Hispanic (using Asian, White, and Other students as the comparison group) were used as the group of risk contributors in the equation because these variables are known to be associated with coping outcomes. Household income was dummy coded 1 for students whose total household income was less than 50% below the Federal Free Lunch guidelines and coded 0 for those whose total household income was 50% or more below the Federal Free Lunch guidelines. We also created dummy variables to represent ethnicity (Black, coded 1 for Black/African American and 0 for all others; Hispanic, coded 1 for Hispanic/Latino and 0 for all others).

As indicated in Table 1, among high-achieving males, risk (i.e., age and being Hispanic) and perceived stress and support (i.e., negative teacher perception, neg-

TABLE 1
Links Between Risk, Net Stress Engagement, and Hypermasculine Attitudes Among High Achieving Adolescent Males

	Model 1[a]		*Model 2*[b]	
	Unstandardized Parameter Estimate	*SE*	*Unstandardized Parameter Estimate*	*SE*
	32.01**	12.01	17.53****	11.95
Risk[c]				
Household income	–2.41	1.91	–1.12	1.69
Age	1.10	0.75	1.63*	0.67
Black	–0.36	2.00	2.28	1.85
Hispanic	–7.35**	2.74	–5.16*	2.49
Net stress engagement[d]				
Negative teacher perception			0.19***	0.06
Negative peer perception			0.21**	0.06
School support			–0.31***	0.09
Support from family, friends, and others			0.03	0.09

[a]Adjusted R^2 = .07. [b]Adjusted R^2 = .32. [c]Partial R^2 = .10.** [d]Partial R^2 = .24.****
*$p < .05$. **$p < .01$. ***$p < .001$. ****$p < .0001$

ative peer perception, and perceived school support) were significantly related to participants' scores on the hypermasculinity index. Specifically, older adolescents were more likely than younger adolescents to endorse stereotypical or hypermasculine sex-role attitudes, and Hispanic/Latino adolescents were less likely than their peers from other ethnic/racial groups to endorse stereotypical or hypermasculine sex-role attitudes. In addition, indexes of stress were also significantly related to exaggerated sex-role attitudes, with high scores on negative teacher perception and negative peer perception each being significantly related to more hypermasculine sex-role attitudes. Finally, high-achieving students who reported feeling more satisfied with the level of emotional support and care they inferred receiving from teachers, principals, and counselors were also less likely to endorse hypermasculine sex-role attitudes than students who reported feeling less satisfied with the level of emotional support received from their teachers, counselors, and principals.

For the regression analyses examining contributors to hypermasculine sex-role attitudes among marginally achieving males, age, household income, and being Black were used as risk variables. As indicated in Table 2, indexes of risk did not contribute significantly to the endorsement of hypermasculine sex-role attitudes. Moreover, similar to findings among high-achieving males, high scores on negative teacher perception and negative peer perception were related to high scores on the hypermasculinity index. Further, adding students' perceptions of emotional support and care from family and friends and from school counselors, teachers, and princi-

TABLE 2
Links Between Risk, Net Stress Engagement, and Hypermasculine Attitudes Among Marginally Achieving Adolescent Males

	Model 1[a]		*Model 2*[b]	
	Unstandardized Parameter Estimate	*SE*	*Unstandardized Parameter Estimate*	*SE*
	62.80	18.0	29.60**	8.79
Risk				
Household income	0.20	2.30	*ns*	
Age	–0.10	1.12	*ns*	
Black	2.20	2.32	*ns*	
Net stress engagement[c]				
Negative teacher perception			0.19*	0.09
Negative peer perception			0.21*	0.09
School support			–0.15	0.09
Support from family, friends, and others			0.11	0.10

[a]Adjusted R^2 = .02. [b]Adjusted R^2 = .19. [c]Partial R^2 = .21.**
*$p < .05$. **$p < .01$.

pals to the regression analysis did not significantly improve the model; however, there was a nonstatistically significant tendency for adolescents who reported greater satisfaction with emotional support and care from teachers, principals, and counselors to also score lower on the hypermasculinity index.

Regression Analyses Exploring the Contribution of Risk and Net Stress Engagement to Positive Attitude Toward Self

As seen in Table 3 among high-achieving males, being Hispanic and Black were significantly related to students' perceived positive attitudes about self, with Hispanics and Blacks reporting higher positive attitudes about self than other groups of students (i.e., Asians, Whites, and Others). In addition, adolescents who scored higher on measures of parental monitoring had more positive self-attitudes than peers who reported lower levels of parental monitoring. Those who reported higher levels of neighborhood risk had significantly lower positive attitudes about self than their peers who perceived less neighborhood-related risk. Although support from family, friends, and others only reached trend level, high-achieving males who reported feeling more emotional support and care from family, friends, and others tended to also have higher positive attitudes about self than those who reported feeling less supported. Among marginally achieving adolescents, the only significant contributor to positive attitudes about self was perceived support

TABLE 3
Links Between Risk, Net Stress Engagement, and Positive Perceptions About Self Among High-Achieving Adolescent Males

	Model 1[a]		*Model 2*[b]	
	Unstandardized Parameter Estimate	*SE*	*Unstandardized Parameter Estimate*	*SE*
	46.25***		31.45**	11.50
Risk[c]				
Household income	0.84	1.65	0.17	1.62
Age	–0.13	0.60	0.15	0.60
Black	8.32***	1.74	7.83***	1.76
Hispanic	7.50**	2.37	6.41**	2.34
Net stress engagement[d]				
Fear of calamity			–0.10*	0.05
Parental monitoring			0.20*	0.08
Support from family, friends, and others			0.14	0.08

[a]Adjusted R^2 = .18. [b]Adjusted R^2 = .25. [c]Partial R^2 = .16****. [d]Partial R^2 = .10**
*$p < .05$. **$p < .01$.

from family, friends, and others. Marginally performing males who reported higher levels of satisfaction with the support they received from significant others were also significantly more likely to have higher scores on positive attitudes toward self (B = .39, *SE* = .09; adjusted $R^2 = .19$, $p < .0001$) than marginally performing males who reported being less satisfied with the support they received from family, friends, and significant others.

DISCUSSION

In this article, we utilized a developmental systems theoretical framework, PVEST, as a dynamic tactic to examine urban male adolescents' lives in context. The predominantly African American sample included both high and low academic achieving adolescents. The two groups did not differ on several demographics, although there was a slight significant difference ($p = .05$) for maternal employment: Marginally performing males, more often than not, had mothers who worked (75%) when compared with high-achieving boys (62%). Similarly, racial and ethnic group membership percentages differed for the high versus low academic performers: Low academic performers were almost twice as likely to be African American. On the other hand, Hispanics and Whites were twice as likely to be high performers versus low performers. Asian Americans, although only 16% of the sample, were over 10 times more likely to be in the high-achieving group. The reasons for these baseline differences were not a subject of this inquiry; however, elsewhere Spencer, Cross, Harpalani, and Goss (2003) discussed reasons for racial disparities in academic achievement.

Our finding that high-performing students perceived receiving significantly more support from school personnel than did low-performing students is both interesting and important because it is generally unclear whether low performance on the part of students generates their perceptions that school support is lacking from school personnel or whether youths' inferred perceptions from school personnel determine low performance. Parallel directional relations were reported by race/ethnicity for the construct perceived negative teacher perceptions. It is impossible to determine whether the low-performing students "earned" the inferred negative teacher perceptions or whether academic performance outcomes were the result of inferred negative teacher perceptions. Of course, social psychology experimental studies by Steele and Aronson (1995) suggested that manipulating environmental stereotypes independent of race/ethnicity has implications for performance outcomes.

The fact that no group differences were apparent for perceptions of parental monitoring; support from family, friends, and others; and fear of calamity aids in "localizing" the apparent "lack of individual and context fit" with regard to school-based experience. The hypermasculinity score, as a reactive coping response, did differentiate high-achieving versus low-achieving students. The finding suggests that low academic performing boys may use bravado behavior to cope with the absence of perceived support from school personnel and, similarly, inferred negative teacher evaluations. Conversely, school personnel, on observing and perhaps fearing perceived hypermasculine behavior and failing to place it in context, may neglect to provide these youth with support. In either case, our emphasis on the responsibility of adults (i.e., school personnel) is deliberate given the fact that urban, adolescent males (and particularly Black males) are often seen through "adultified" lenses rather than being viewed as youth developing in what are often high-risk contexts.

The pattern of findings supports the view that besides curriculum improvements, additional tasks of whole school reform efforts should center on improvement of the quality of school personnel–student relationships (i.e., perceptions of mutual valuing, trust, and respect). In fact, consistent with the themes undergirding Steele and Aronson's (1995) social psychology experimental studies, the findings that specific youth demographics predicted heightened hypermasculinity suggest possible stereotyping. Specifically, age (being older), extremely low household income, and ethnicity function as risks associated with specific coping methods: hypermasculine attitudes and low positive attitudes about self. Similarly, as a system of adaptations, low positive self-attitudes reinforce the deployment of reactive hypermasculine coping methods.

The pattern of results also suggests that more research is needed on the impact of perceptions concerning a sense of calamity (i.e., fear of calamity score) and positive perceptions about the self. For example, the finding that for high-performing males, fear of calamity reached trend level only when associated with less positive self-attitudes is both interesting and informative. The relation implies that psychological preoccupation with net stress (i.e., a state of fearfulness) can interfere with academic persistence and suggests policy and training implications. Consistent with Maslow's (1967) hierarchy of basic needs required for survival (e.g., need for warmth, food, shelter, protection), it is critical to remember the foundational relation between satisfaction of basic needs (e.g., a sense of belonging) and higher order psychological functioning (e.g., intellectual pursuits). We believe it is important to note the apparent inverse relation between the availability of cognitive energy for academic challenges and the presence of heightened emotional stressors in competition for the jointly shared psychic energy. In fact, this is consistent with the idea that there is finite psychic energy accessible at any given time; this energy needs to be both adequate and obtainable to both cognitive and emotional domains of functioning. Our study design did not provide an opportunity for a direct test of this view, but it is suggestive of important future work. Furthermore, the finding that marginally performing male students who reported higher levels of inferred support also obtained scores indicating more positive attitudes about the self (as inferred from family, friends, and others) provides additional support for the given interpretation.

Also important to note is the possible dissonant character of adaptive behaviors across contexts that youth encounter, as mentioned in our overview (e.g., Stevenson, 1997). Phelan, Davidson, and Cao (1991) discussed how the transitions between settings can be difficult to negotiate for many vulnerable youth, and this issue was covered in more detail elsewhere (Spencer, Harpalani, Fegley, Dell'Angelo, & Seaton, 2003). In the wider scope, hypermasculine attitudes may emerge as a result of particular contextual risks, but they are also related to an array of broader social inequities and to unhealthy (although normative) notions of masculinity in America. Addressing these factors adequately will require broad societal changes in the structure, composition, and resources of urban neighborhoods and the socialized gender norms of American society. Although it is important to work toward such changes, one must also acknowledge that they will come about slowly. In the meantime, one should recognize the various factors that may lead to coping responses such as hypermasculinity and incorporate this knowledge into smaller scale policies, professional training standards, and interventions that may mitigate risk and promote resilience.

In this vein, considered together, the findings suggest multiple strategies for interventions and supportive programming for families supportive of cultural socialization and innovations for school personnel training and evaluative oversight. If having a positive attitude is a significant contributor to diminished levels of reactive coping (i.e., hypermasculinity), strategies to enhance the relation between home and school, particularly for low performing African American male students, should go a long way toward assisting such youth to view school as a psychologically safe setting and a trustworthy learning environment. As has been suggested by Erikson (1966) and other theorists, trust is the first important developmental task confronted in the first year of life and on which all learning and normative developmental processes are structured. Trust has an important role for academic skill enhancement; learning often requires that youth acknowledge what they don't know in order to learn what they need to learn for a particular developmental period. This imperative exists independently of the specific tasks confronted, and its variability may be due to individual differences in the presence of protective factors (e.g., those with the greatest home–school similarity). Trust does not develop if animosity and particular reactive coping responses, which may include expressions of hypermasculinity, are allowed to fester without the availability of positive and proactive models of adaptive coping, which are necessary for resilience. In fact, maladaptive coping responses such as hypermasculinity may be defensive reactions to a lack of trust; they can serve an immediate, protective function in this manner. However, it is important to recognize that such expressions can mask vulnerability, which would more likely be expressed if trust were present.

If provided with proper training and oversight, teachers and school personnel can proactively facilitate students' trust—by creating school contexts that youth perceive as psychologically and physically safe. If the appropriate professional supports are available, such as informed training and evaluative oversight to teachers and personnel, this also enhances school personnel's trust in the educational system and helps to meet adults' own developmental needs (i.e., an accrued efficacious professional self). Thus, by creating a trusting environment that is perceived as safe and supportive by youth, schools can facilitate adaptive coping responses for both urban students and school personnel.

Finally, research that makes use of a more dynamic framework would aid in designing both teacher training programming and its evaluation. Our pattern of findings proposes that a phenomenological theoretical perspective is of assistance as suggested by data interpretations that indicate the importance of cognition determining inference-making processes. The use of a conceptual framework that acknowledges the motivational link between reactive behavior, identity processes, and academic performance as coping outcomes provides a useful heuristic device for understanding the complexities inherent in such an individual-context analysis. It is in this light that we employ PVEST as a model to understand risk and resilience from a normative human development perspective.

ACKNOWLEDGMENTS

This research was funded through grants to Margaret Beale Spencer from the National Institutes of Mental Health, Office of Educational Research and Improvement, and the Ford and Kellogg Foundations. We acknowledge Farrah Samuels and Janine Beaty-Hunter for their excellent service in programming activities related to this research project.

REFERENCES

Abbott, A. A. (1981). Factors related to third grade achievement: Self-perception, classroom composition, sex, and race. *Contemporary Educational Psychology, 6*, 167–179.

Block, J. (1985, October). *Some relationships regarding the self emanating from the Block and Block longitudinal study*. Paper presented at the Social Science Research Council conference, Center for Advanced Study in the Behavioral Sciences, Stanford, CA.

Bronfenbrenner, U. (1989). Ecological systems theory. In R. Vasta (Ed.), *Annals of child development* (pp. 187–248). Greenwich, CT: JAI.

Connell, J. P., Spencer, M. B., & Aber, J. L. (1994). Educational risk and resilience in African American youth: Context, self, action and outcomes in school. *Child Development, 65*, 493–506.

Cooley, C. H. (1902). *Human nature and the social order*. New York: Scribner's.

Cunningham, M. (1994). *Expressions of manhood: Predictors of educational achievement and African American adolescent males (boys)*. Unpublished doctoral dissertation, Emory University, Atlanta, GA.

Cunningham, M. (1999). African American adolescent males' perceptions of their community resources and constraints: A longitudinal analysis. *Journal of Community Psychology, 27*, 569–588.

Donziger, S. (Ed.). (1996). *The real war on crime*. New York: Harper Perennial.

D'Souza, D. (1995). *The end of racism: Principles for a multiracial society*. New York: Free Press.

Erikson, E. (1966). The concept of identity in race relations: Notes and queries. *Daedalus, 95*, 145–171.

Erikson, E. (1968). *Identity: Youth and crisis*. New York: Norton.

Ferguson, A. A. (2001). *Bad boys: Public schools and the making of Black masculinity*. Ann Arbor: University of Michigan Press.

Ferraro, K. (1995). *Fear of crime: Interpreting victimization risk*. Albany: State University of New York Press.

Glassner, B. (1999). *The culture of fear: Why Americans are afraid of the wrong things*. New York: Basic Books.

Goodey, J. (1997). Boys don't cry: Masculinities, fear of crime and fearlessness. *British Journal of Criminology, 37*, 401–418.

Hare, B. R. (1977). Racial and socioeconomic variation in preadolescent area-specific and general self-esteem. *International Journal of Intercultural Relations, 3*, 31–51.

Hare, B. R., & Castenell, L. A., Jr. (1985). No place to run, no place to hide: Comparative status and future prospects of Black boys. In M. B. Spencer, G. K. Brookins, & W. R. Allen (Eds.), *Beginnings: The social and affective development of Black children* (pp. 201–214). Hillsdale, NJ: Lawrence Erlbaum Associates, Inc.

Kaufman, P., Chen, X., Choy, S. P., Ruddy, S. A., Miller, A. K., Fleury, J. K., Chandler, K. A., Rand, M. R., Klaus, P., & Planty, M. G. (2000). *Indicators of school crime and safety, 2000* (NCES Publication No. 2001–017/NCJ–184176). Washington, DC: U.S. Departments of Education and Justice.

Luthar, S. (Ed.). (2003). *Resilience and vulnerability: Adaptation in the context of childhood adversities*. New York: Cambridge University Press.

Luthar, S. S., & Cicchetti, D. (2000). The construction of resilience: Implications for interventions and social policies. *Development & Psychopathology, 12*, 857–885.

Luthar, S. S., Cicchetti, D., & Becker, B. (2000). The construct of resilience: A critical evaluation and guidelines for future work. *Child Development, 71*, 543–562.

Maslow, A. H. (1967). *Motivation and personality* (3rd ed.). New York: Harper & Row.

May, D. (2001). *Adolescent fear of crime, perceptions of risk, and defensive behaviors: An alternative explanation violent delinquency*. Lewiston, NY: Edwin Mellon Press.

McDermott, P. A., & Spencer, M. B. (1995a). *Measurement Properties of Hare/Funder/Block Ego-Esteem/Resilience Scale (Years 2 and 3)* in (Interim Research Report No. 8). Philadelphia, PA: University of Pennsylvania, Center for Health, Achievement, Neighborhood, Growth, and Ethnic Studies.

McDermott, P. A., & Spencer, M. B. (1995b). *Measurement Properties of Revised Student Perceived Parental Monitoring* in (Interim Research Report No. 10). Philadelphia, PA: University of Pennsylvania, Center for Health, Achievement, Neighborhood, Growth, and Ethnic Studies.

Mosher, D. L., & Sirkin, M. (1984). Measuring a macho personality constellation. *Journal of Research and Personality, 18*, 150–163.

Moynihan, D. (1965). *The Negro family: The case for national action*. Washington, DC: U.S. Department of Labor.

Noll, E., Cassidy, E. F., & Spencer, M. B. (2004). *Monetary incentive use and achievement among adolescents from low-resource backgrounds*. Manuscript submitted for publication.

Phelan, P., Davidson, A. L., & Cao, H. T. (1991). Students' multiple worlds: Negotiating the boundaries of family, peer, and school cultures. *Anthropology & Education Quarterly, 22*, 224–250.

Riechard, D. E., & McGarrity, J. (1994). Early adolescents' perceptions of relative risk from 10 societal and environmental hazards. *Journal of Environmental Education, 26*, 16–23.

Sampson, R. (1986). Effects of socioeconomic context on official reaction to juvenile delinquency. *American Sociological Review, 51*, 876–885.

Sampson, R., & Raudenbush, S. (1999). Systematic social observation of public spaces: A new look at disorder in urban neighborhoods. *American Journal of Sociology, 105*, 603–651.

Sarason, I. G., Levine, H. M., Basham, R. B., & Sarason, B. R. (1983). Assessing social support: The social support questionnaire. *Journal of Personality and Social Psychology, 44*, 127–139.

Shaw, C., & McKay, H. (1946). *Juvenile delinquency in urban areas*. Chicago: University of Chicago Press.

Shoemaker, A. L. (1980). Construct validity of area specific self-esteem: The Hare Self-Esteem Scale. *Educational and Psychological Measurement, 40*, 495–501.

Snell, C. (2001). *Neighborhood structure, crime, and fear of crime: Testing Bursik and Gramsick's neighborhood control theory*. New York: LLB Scholarly Publishing.

Spencer, M. B. (1995) Old issues and new theorizing about African American youth: A Phenomenological variant of ecological systems theory. In R. L. Taylor (Ed.), *African-American youth: Their social and economic status in the United States* (pp. 37–69). Westport, CT: Praeger.

Spencer, M. B. (1999). Social and cultural influences on school adjustment: The application of an identity-focused cultural ecological perspective. *Educational Psychologist, 34*, 43–57.

Spencer, M. B. (2000). Identity, achievement, orientation, and race: "Lessons learned" about the normative developmental experiences of African American males. In W. H. Watkins, J. H. Lewis, & V. Chou (Eds.), *Race and education: The role of history and society in the education of African American students* (pp. 100–127). Boston: Allyn & Bacon.

Spencer, M. B. (2001). Resiliency and fragility factors associated with the contextual experiences of low resource urban African American male youth and families. In A. Booth & A. C. Crouter (Eds.), *Does it take a village?* (pp. 51–77). Mahwah, NJ: Lawrence Erlbaum Associates, Inc.

Spencer, M. B., Cross, W. E., Harpalani, V., & Goss, T. N. (2003). Historical and developmental perspectives on Black academic achievement: Debunking the "acting White" myth and posing new directions for research. In C. C. Yeakey & R. D. Henderson (Eds.), *Surmounting all odds: Education, opportunity, and society in the new millennium* (pp. 273–304). Greenwich, CT: Information Age Publishers.

Spencer, M. B., Dupree, D., & Swanson, D. P. (1996). Parental monitoring and adolescents' sense of responsibility for their own learning: An examination of sex differences. *Journal of Negro Education, 65*, 30–42.

Spencer, M. B., Fegley, S. G., & Harpalani, V. (2003). A theoretical and empirical examination of identity as coping: Linking coping resources to the self processes of African American youth. *Applied Developmental Science, 7*, 181–188.

Spencer, M. B., & Harpalani, V. (2001). African American adolescents, research on. In R. M. Lerner & J. V. Lerner (Eds.), *Adolescence in America: An encyclopedia* (Vol. 1, pp. 30–32). Denver, CO: ABC-CLIO.

Spencer, M. B., Harpalani, V., Fegley, S., Dell'Angelo, T., & Seaton, G. (2003). Identity, self, and peers in context: A culturally-sensitive, developmental framework for analysis. In R. M. Lerner, F. Jacobs, & D. Wertlieb (Eds.), *Handbook of applied developmental science: Promoting positive child, adolescent, and family development through research, policies, and programs* (Vol. 1, pp. 123–142). Thousand Oaks, CA: Sage.

Spencer, M. B., Noll, E., Stoltzfus, J., & Harpalani, V. (2001). Identity and school adjustment: Questioning the "acting White" assumption. *Educational Psychologist, 36*, 21–30.

Steele, C., & Aronson, J. (1995). Stereotype threat and the intellectual test performance of African Americans. *Journal of Personality and Social Psychology, 69*, 797–811.

Stevenson, H. C. (1997). Missed, dissed, and pissed: Making meaning of neighborhood risk, fear and anger management in urban Black youth. *Cultural Diversity and Mental Health, 3*(1), 37–52.

Thoits, P. (1995). Stress, coping, and social support processes: Where are we? What next? *Journal of Health and Social Behavior, 35*, 53–79.

Welsh, W., Stokes, R., & Greene, J. (2000). A macro-level model of school disorder. *Journal of Research in Crime and Delinquency, 37*, 243–283.

Wilson, W. J. (1996). *When work disappears: The world of the new urban poor*. New York: Vintage.

RESEARCH IN HUMAN DEVELOPMENT, *1*(4), 259–290

Traumatic Events, Psychiatric Disorders, and Pathways of Risk and Resilience During the Transition to Adulthood

Betsy J. Feldman and Rand D. Conger
University of California, Davis

Rebecca G. Burzette
Iowa State University

The goal in this study was to investigate the occurrence, frequency, and consequences of traumatic events during the first 2 decades of life. The sample comprised 524 Euro-American young adults from the rural Midwest who are part of an ongoing longitudinal study (see Conger & Conger, 2002). We found that psychiatric disorders were associated with traumatic events and particularly strongly associated with experiences of childhood maltreatment. Maltreated children were 5.64 times more likely to develop a disorder than individuals who experienced no trauma. These odds varied, depending on the disorder, from more than 3 for alcohol dependence to more than 18 for drug dependence. We found that social support acted to reduce risk for emotional disorder and buffered effects of traumatic events under some circumstances, especially in cases of childhood trauma.

Traumatic events are part of human existence; unavoidable, unfortunate, often unpredictable, but not uncommon (Breslau et al., 1998; Kessler, 2000; Resnick, Kilpatrick, Dansky, Saunders, & Best, 1993). In addition to the immediate effects of these events, a significant proportion of those who suffer serious traumas continue to experience emotional aftershocks that may persist well into the future (Breslau et al., 1998; Kessler, Sonnega, Bromet, Hughes, & Nelson, 1995). These emotional aftereffects, in the form of acute stress disorder, posttraumatic stress disorder (PTSD), and other emotional problems (Kessler et al., 1995; Lipschitz,

Requests for reprints should be sent to Betsy J. Feldman, Family Research Group, 202 Cousteau Place, Suite 100, Davis, CA 95616. E-mail: bfeldman@ucdavis.edu

Rasmusson, & Southwick, 1998; Pine & Cohen, 2002; Tiet et al., 2001) may be of special concern when they occur in the young. Because of their emotional and biological immaturity, children and adolescents are especially vulnerable not only to the immediate effects of trauma (e.g., Pine & Cohen, 2002; Pynoos, Steinberg, & Piacentini, 1999) but also to potential long-term effects (e.g., Lipschitz et al., 1998; Rutter & Maughan, 1997). In this article, we examine the prevalence of trauma and the influence of traumatic life events on emotional and behavioral problems across the years of adolescence and early adulthood.

HOW COMMON IS TRAUMA IN CHILDHOOD AND ADOLESCENCE?

For purposes of this study, we are concerned with serious traumas, which have been defined as events "outside the range of usual human experience and that would be markedly distressing to almost anyone" (*Diagnostic and Statistical Manual of Mental Disorders*, 3rd ed., revised [*DSM–III–R*]; American Psychiatric Association, 1987, p. 250). The risk that a young person will experience a traumatic event serious enough to potentially qualify for a diagnosis of PTSD depends on many things, but to a large extent it is dependent on their surroundings (Kessler, 2000). In the United States, for example, urban youth are at greater risk of exposure to violence than those who live in rural or suburban areas (Singer, Anglin, Song, & Lunghofer, 1995), whereas in less developed countries, youth are at greater risk for a variety of types of interpersonal violence including not only crime but also political violence (Kessler, 2000).

Although several researchers have studied the prevalence of trauma in adult populations (e.g., Kessler et al., 1999; Norris, 1992), just a few have investigated the prevalence of trauma in general in community samples of children or adolescents, although certain specific traumatic events, such as child sexual abuse and urban violence, have been examined more thoroughly. Three studies (Costello, Erkanli, Fairbank, & Angold, 2002; Elklit, 2002; Giaconia et al., 1995) have found that the risk for experiencing one or more traumatic events ranged from 25% (Costello et al., 2002) to 88% (Elklit, 2002) among adolescents. Differences in methodology and ages may account for differences in the prevalence of trauma. Elklit (2002), for instance, inquired about events not included on other researchers' lists, such as the death of somebody close (the most frequently reported) and the absence of a parent. The longer list of traumas likely accounted for the high prevalence of negative events in this study. Costello et al.'s (2002) sample, on the other hand, included younger adolescents (13–16 years), and their relative youth may have reduced opportunities for exposure to trauma, which likely increases during late adolescence. For example, 43% of Giaconia et al.'s (1995) sample of 18-year-old high school seniors reported having experienced one or more traumas. From these studies, we conclude that on average, adolescents appear to experience a significant number of traumas in their lives and that the risk for trauma likely increases with age.

WHAT ARE THE AFTEREFFECTS OF TRAUMA?

Although traumatic events have been related to several types of psychiatric disorder (Goenjian et al., 1995; Heim & Nemeroff, 2001; Kessler et al., 1995; Kilpatrick et al., 2003; Pine & Cohen, 2002), PTSD is historically the primary disorder specifically linked to exposure to a serious trauma. Symptoms of PTSD are grouped into three categories: (a) reexperiencing stressful situations such as in dreams or, in children, repeated traumatic play; (b) avoidance of reminders of the trauma and psychological "numbing," or dampening of affect; and (c) ongoing heightened arousal (*DSM*, 4th ed., text revision; American Psychiatric Association, 2000). In children, symptoms may also involve academic problems, behavior and conduct problems, sleep disturbances, and difficulties in social interactions with peers and family (Pynoos, Steinberg, & Wraith, 1995). Some traumas are more consistently linked to PTSD than others. Among adults in the United States, sexual assaults (Nishith, Mechanic, & Resick, 2000; Ullman & Filipas, 2001) and combat and prisoner-of-war experiences (Benotsch et al., 2000; Ginzburg, Solomon, Dekel, & Neria, 2003; King, King, Foy, & Gudanowski, 1996; Macklin et al., 1998; Sutker, Davis, Uddo, & Ditta, 1995) create the greatest risk for PTSD among victims. Among children, studies have found a dose–response relation between exposure to trauma and severity of response (Duraković-Belko, Kulenović, & Đapić, 2003; Fergusson & Lynskey, 1997; Goenjian et al., 1995; Green et al., 1991; Zoroglu et al., 2003) such that more severe traumas—those that cause greater harm or that are associated with greater loss (Pynoos et al., 1995)—are more likely to result in disorder.

For children and adults, other aspects of the trauma and aspects of the individual can increase or decrease the risk of developing PTSD or other pathological outcomes. The duration and degree of exposure to danger are related to the risk for PTSD (Duraković-Belko et al., 2003; Fergusson, Horwood, & Lynskey, 1996; Goenjian et al., 1995; Pine & Cohen, 2002) as are gender (e.g., Bolton, O'Ryan, Udwin, Boyle, & Yule, 2000; Giaconia et al., 1995) and, especially in children and adolescents, the age of the victim (e.g., Fletcher, 2003; Pynoos et al., 1995). Women and girls are usually found to be at greater risk for negative outcomes following traumas, even when considering the same types of traumas for males and females (e.g., Pine & Cohen, 2002; Singer et al., 1995; Springer & Padgett, 2000; but see also Elbedour, ten Bensel, & Bastien, 1993; Rossman, Bingham, & Emde, 1997). Women and girls are at greater risk for sexual assault, which is one of the traumatic events most likely to cause serious and lengthy psychological disorders

(Brewin, Andrews, & Valentine, 2000). In addition, they are more likely to experience maltreatment, including sexual abuse, in childhood (Costello et al., 2002; Kessler et al., 1995; Singer et al., 1995), which also carries particular risk for serious long-term harm.

The Special Vulnerability of Children and Adolescents

Because of their stage of emotional and physical development, children and adolescents are especially vulnerable to the effects of trauma (Pynoos et al., 1995). Many researchers have found that trauma at a young age can negatively affect later emotional development (Brown, Cohen, Johnson, & Salzinger, 1998; Brown, Cohen, Johnson, & Smailes, 1999; Farley & Patsalides, 2001; Rutter & Maughan, 1997; Tyano et al., 1996). Children are still forming expectations of the outside world, and traumatic events can interfere with the development of their basic senses of trust and personal safety as well as their expectations of the world and themselves (Fletcher, 2003; Pynoos et al., 1999).

Depending on their stage of development, children may be dependent on parents and other adults to help them cope with and understand feelings and emotions that arise during and after traumatic events, so their surroundings at the time of and following the event are critical to their ability to process their emotional and physiological reactions (Pine & Cohen, 2002; Pynoos et al., 1995). Therefore, it is no surprise that events can have particularly negative and lasting effects when they involve deprivation of parental support, as they would if family members are lost (e.g., war-related events or natural disasters), if parents cause the event (e.g., maltreatment), or when the parents themselves are unable to cope with events due to their own vulnerability or psychopathology (Bolton et al., 2000; Pynoos et al., 1995). In addition to these emotional dependencies and vulnerabilities, children are more susceptible than adults to physiological changes in brain structure that may follow traumatic events and can create a greater risk of future disorders (Heim & Nemeroff, 2001; Lipschitz et al., 1998).

A multitude of studies have confirmed that children who have experienced traumatic events of various sorts are more likely than those who have not to develop several different types of psychiatric disorder including PTSD. To a significant degree, the form of psychopathology depends on age and the type of trauma (Fletcher, 2003; Korol, Kramer, Grace, & Green, 2002; Lipschitz et al., 1998; Pine & Cohen, 2002; Pynoos et al., 1995; Wright, Masten, & Hubbard, 1997); however, problems with anxiety and depression are common under most circumstances (Finkelhor & Kendall-Tackett, 1997; Heim & Nemeroff, 2001; Pine & Cohen, 2002; Tiet et al., 2001). In addition, many children, although more commonly boys, will demonstrate conduct problems and acting-out behaviors (Dodge, Pettit, & Bates, 1997; Fergusson et al., 1996; Kessler et al., 1995; Tiet et al., 2001). Later, in adolescence and adulthood, childhood trauma continues to be associated with greater likelihood for several forms of psychopathology including depression and anxiety (Edwards, Holden, Felitti, & Anda, 2003; Fergusson et al., 1996; Goenjian et al., 1995; Kessler et al., 1995; Pine & Cohen, 2002), substance abuse or dependence (Fergusson et al., 1996; Giaconia et al., 1995; Kessler et al., 1995), borderline personality disorder (Zanarini & Frankenburg, 1997), and antisocial personality disorder (Trickett, Reiffman, Horowitz, & Putnam, 1997).

In addition, children who experience significant trauma may suffer from long-term problems with mastery and self-esteem (Pynoos et al., 1995; Wright et al., 1997), medical problems (Farley & Patsalides, 2001), and a generally lower sense of well-being (Pynoos et al., 1995; Royse, Rompf, & Dhooper, 1991). Some childhood traumas are also associated with later suicide ideation and attempts (Fergusson et al., 1996; Zoroglu et al., 2003) and self-mutilation (Zoroglu et al., 2003).

Also important, many childhood traumas do not occur in isolation. There are frequently other circumstances, such as discordant households and community violence (Brown et al., 1999; Cicchetti, Toth, & Lynch, 1997; Dodge et al., 1997; Osofsky & Scheeringa, 1997), which may also contribute to the disorders that follow the traumatic experiences. In addition, children often have more than one negative experience at a time (e.g., abuse and neglect; Edwards et al., 2003; Egeland, 1997). Moreover, there are differences between children that some speculate may make them more likely to encounter some types of traumatic events (Brown et al., 1998; Dodge et al., 1997; Osofsky & Scheeringa, 1997; Wright et al., 1997). However, exposure to trauma can make a child more sensitive to stressors and more irritable and difficult for parents and others to deal with. Thus, traumatic experiences may lead to changes in a child's temperament or personality, which may then lead to a greater likelihood of further traumatization (Pynoos et al., 1995; Wright et al., 1997).

Protective Influences

In terms of both improved theoretical understanding and the development of more effective interventions, an important consideration concerns processes or mechanisms that protect children from psychopathological responses to traumatic events. A number of outside factors may put children at greater or lesser risk for poor outcomes. The response of parents, for example, may have a substantial influence on the child's ability to recover from a trauma. Parent psychopathology prior to or following a trauma may increase the risk of child disorder (Goenjian et al., 1995; Green et al., 1991; Pynoos et al., 1999; Saigh, Yasik, Sack, & Koplewicz, 1999), whereas warm, supportive, competent parenting and a calm household may help a child recover (Fletcher, 2003; Pynoos et al., 1995).

Separation from family and the loss of other social supports, such as those provided by school and friends, may worsen the effects of a traumatic event

(Goenjian et al., 1995; Pine & Cohen, 2002; Pynoos et al., 1999, 1995). On the other hand, the presence and reliable continuity of social support is hypothesized to have important protective effects (Pine & Cohen, 2002; Turner, 1999; Wolfe & Wekerle, 1997; Wright et al., 1997). Social support may not always be helpful, however. If the victim of trauma does not have a sense of being supported or helped, then he or she may not benefit as much (Sarason, Pierce, & Sarason, 1994; Turner, 1999). Also, if a relationship in which support is offered is characterized by conflict, the support may not be helpful to the victim (Sarason et al., 1994). Thus, for instance, parental support offered in a family in which family communication is poor (as might be the case when abuse has been present) or in which parent–child strife is common may not be accepted by the child or simply may not be perceived.

THIS STUDY

The studies just reviewed indicate that different kinds of traumatic events may have different consequences and also that the same types of events may have different consequences depending on the circumstances surrounding the individual. In this study, we investigated a range of serious traumas that may occur during the period from childhood to the early adult years. We also examined specific aspects of the respondent's environment that may alter the influence of these traumas on risk for psychiatric problems. Our goals for studying trauma in this sample were twofold: (a) we wanted to know what traumatic events were likely to occur in the lives of children and adolescents who grew up in the rural Midwest between the 1970s and the 1990s, and (b) we sought to discover how the victims of trauma fared in early adulthood in terms of their mental health. In this study, we used a combination of methodologies to address these issues. We employed in-depth, multiinformant data from a rural community cohort of adolescents beginning when they were in ninth grade to assess their environment, traumatic experiences, and disorders.

Consistent with earlier research, we expected to find that exposure to trauma would be positively associated with mental health difficulties. We also expected that high levels of social support from inside and outside the family would reduce the likelihood of psychiatric disorder through two complementary processes. First, Masten (2001) proposed that resilience-promoting mechanisms such as social support serve to compensate for stress and trauma by decreasing distress across all levels of trauma, a statistical main effect. Second, we expected that social support would buffer the impact of trauma for those individuals who were at highest risk by virtue of exposure to traumatic events, a statistical interaction or moderating effect (Conger & Conger, 2002; Turner,

1999). The longitudinal design of the study made it possible to examine these hypothesized processes of risk and resilience as they played out across the years of adolescence and early adulthood.

METHOD

Participants

Data come from the Family Transitions Project, a longitudinal, community epidemiological study of 558 target youth, their families, and selected close relationships. Data collection of various types began when they were in either the seventh (1989) or ninth (1991) grade and continued on an annual basis through the 1990s with a 90% retention rate through 1999. Four hundred fifty-one of the target youth came from two-parent families, and 107 were part of a matched sample from single-parent, mother-headed households. The target youth from two- and single-parent families were matched in terms of grade level and geographic location. The target youth, their parents, and a close-aged sibling participated in the study. Because of a very small minority population in rural Iowa, where the research was initiated, all participants are of European heritage. Participants in the study were originally recruited to examine the family and developmental effects of the economic downturn in agriculture of the 1980s; for that reason, they were recruited from eight rural counties in Iowa. The original sample of families was primarily lower middle class or middle class. Additional details about the study can be found in Conger and Conger (2002), Conger and Elder (1994), and Simons and Associates (1996).

Procedures

During the adolescent years, target youth and their families were recruited from public and private schools in rural areas of Iowa. Eligible families were first sent a letter explaining the project and then were contacted via telephone and asked to participate. Over 90% of the eligible single-parent families and 78% of the eligible two-parent families agreed to be interviewed. The higher recruitment rate for single-mother headed families appeared to result from the greater ease in scheduling visits when only three as opposed to four family members are asked to participate and from the tendency of mothers to be more interested than fathers in being involved in this type of research.

Beginning as early as 1989 (seventh grade), the target adolescents and their families were visited twice each year in their homes by a trained interviewer. In this article, we focus on adolescent data collected in 1991 (ninth grade), the earliest point in time when data are available from the full sample of single- and two-parent families. The target adolescents, in this visit, ranged in age from 14 to 17 years ($M = 15.44$, $SD = 0.54$). Each visit lasted approximately 2 hr, with the second visit occurring within 2 weeks of the first visit. During the first visit, participating family members individually completed questionnaires pertaining to subjects such as the demographic characteristics of the family and the personal characteristics of themselves and other family members. During the second visit, family members participated in four videotaped interaction tasks. Two of these tasks were used in these analyses: the family problem-solving task (15 min) and the sibling interaction task (15 min), which followed the problem-solving interaction. These tasks were designed to elicit a broad range of family member behaviors. Details regarding the interaction tasks used in these analyses are provided in the following section on study measures. However, it is important to note that a common element of these tasks is an attempt to generate a challenging or stressful situation for family members. In the face of this type of challenge, we found that family members are likely to demonstrate a range of emotions that can range from warmth and support to anger and hostility (Melby & Conger, 2001). Trained observers coded the quality of family interactions using the Iowa Family Interaction Rating Scales (Melby & Conger, 2001), which have demonstrated adequate reliability and validity.

In 1995, 1997, and 1999, after our participants graduated from high school, a structured diagnostic interview was administered in the target youth's place of residence to evaluate psychiatric disorder. The 1st year the diagnostic interview was administered, 1995, was the year after most participants graduated from high school. Their ages ranged from 18 to 21 years ($M = 19.15$, $SD = 0.42$). In 1997 and 1999, lifetime traumatic exposure was assessed. In 1997, the participants were asked about lifetime traumatic events, and during the 1999 session, the questions referred to the intervening years. Participants who were not available to be interviewed in 1997 were asked the lifetime questions in 1999.

Measures

Control variables. Family characteristics, such as lower income and education, have also been associated with psychiatric disorder in children (Sameroff, 2000). Two variables were included to control for these possible influences. We used the parents' education measured in years of education and averaged across the two parents in cases in which there were two parents and the family's income-to-needs ratio calculated from their 1991 economic report data. Income to needs reflects total household income divided by the poverty line for a family of a given size.

Psychiatric diagnoses and traumatic events. Psychiatric disorders and traumatic experiences were assessed using the University of Michigan modification of the Composite International Diagnostic Interview (UM–CIDI; Kessler et al., 1994). This fully structured diagnostic interview generates estimates of *DSM–III–R* (American Psychiatric Association, 1987) psychiatric disorders in terms of both onsets and recurrences for adolescents and adults. World Health Organization field trials showed that the CIDI possesses good reliability and validity (Wittchen, 1994). To further assure reliability, all interviewers underwent a 5-day training workshop. Interview materials were double-checked by research staff, and all interviews were audiotaped. Reliability tests using 10% of the audiotapes selected at random showed 100% agreement between the interviewers and advanced counseling psychology graduate students using the CIDI interview form. The diagnoses assessed in the UM–CIDI interview include major depression, mania, dysthymia, simple and social phobias, panic disorder and agoraphobia, generalized anxiety disorder, PTSD, alcohol and drug abuse and dependence, conduct disorder, and antisocial personality disorder. Excluded from this study are rare diagnoses, such as somatization disorder, and disorders that require more extended clinical interviews for reliable diagnoses such as schizophrenia. In the following analyses, we primarily predict risk for categories of disorder (e.g., any disorder, affective disorders, etc.). Trauma was measured using the UM–CIDI list of traumatic events considered severe enough to lead to an evaluation of PTSD.

Social support in the family. One composite variable from 1991 (ninth grade) was used to evaluate targets' social support within the family. The within-family support variable comprised a combination of the target youth's reports on the nature of his or her interactions with family members in the study (parents and one sibling) and observers' ratings of videotaped interactions with the same family members (the single-parent-family interactions did not include the father). Several steps were taken to compose the overall family support variable. First, all items were recoded as needed so that higher numbers indicated more supportive relationships or interactions. Next, each subscale was averaged and then standardized. Finally, the subscales were averaged together and the resulting variable was standardized for ease of interpretation in the analyses. All subscale means, standard deviations, and alpha coefficients are reported in Table 1. The alpha coefficient for the 11 subscales that made up the family support variable is .86.

Three sets of questions produced a total of nine subscales that were used to assess the target adolescent's perception of supportiveness in his or her relationship with each family member (each parent and the participating sibling). Subscales related to relationship quality, positive versus negative interactions, and identification with parents and siblings were generated from target report (Table 1). Target report of relationship quality comprised a set of three questions that assessed

TABLE 1
Descriptive Statistics for Social Support Measures (1991, Ninth Grade)

Source	*Measure*	*M*	*SD*	*No. Items or Subscales*	α
Family support measures					
Target report					
Father	Quality	2.92	0.70	3	.82
	Positive vs. Negative	5.62	0.94	24	.95
	Identification or Rejection	4.03	0.76	9	.92
Total	All Items			36	.96
Mother	Quality	3.16	0.60	3	.73
	Positive vs. Negative	5.63	0.81	24	.94
	Identification or Rejection	4.04	0.64	9	.89
Total	All Items			36	.95
Sibling	Quality	3.04	0.67	3	.77
	Positive vs. Negative	4.46	1.05	24	.95
	Identification and Closeness	2.85	0.49	23	.90
Total	All Items			50	.96
Target-report total	Subscales			9	.86
Family support measures					
Observer ratings					
Positive	Father	2.09	0.53	5	.76
	Mother	2.23	0.55	5	.77
	Sibling	2.72	0.80	5	.89
Total		2.38	0.51	15	.84
Negative (lower is more negative)	Father	3.69	0.91	3	.85
	Mother	3.45	0.98	3	.88
	Sibling	3.17	1.22	3	.92
Total		3.40	0.79	9	.83
Observer ratings total	All Items			24	.89
Family support total	All Subscales			11	.86
Support from friends	Perceived Support	3.33	0.40	11	.74
	Direct Support	4.36	0.53	5	.76
Total	All Items			16	.82
Support from school	School Support	3.73	0.63	10	.85
	School Attachment	3.74	0.61	10	.84
	School Problems	3.03	0.52	14	.88
Total	All Items			34	.90

the adolescent's overall satisfaction, happiness, and ability to communicate with each family member. Four possible levels of answers were available from "very [satisfied]" or [talks about private things] "a lot" to "very un[satisfied]" or [talks about private things] "not at all." The alphas for these measures of relationship quality ranged from .73 (mother) to .82 (father; see Table 1).

A second set of questions tapped into the acts of warmth and social support versus expressions of hostility and anger in the relationships during the past month. To evaluate warmth and social support, the target youths responded to questions such as "How often in the past month did [family member] let you know he or she cares about you?" Hostility and coercion were assessed by questions like "How often in the past month did [family member] hit, push, grab, or shove you?" Questions could be answered on a 7-point scale ranging from 1 (*always*) to 7 (*never*). Theoretically, we proposed that the most supportive relationships would be ones that were high on care and affection and low on hostility and coercion. We also assumed that affection and hostility would represent opposite ends of a single dimension of supportiveness. High alpha coefficients (.94–.95) for these scales supported this assumption (see Table 1).

The questions in the third set (the Identification subscale) differed depending on whether they referred to the parents or sibling. In the case of parents, the questions tapped into parents' acceptance versus rejection of the adolescent and the adolescent's identification with his or her parents, a sign of closeness and warmth in their relationship (Elder & Conger, 2000). This set of items includes mostly positive statements such as "I have a lot of respect for my [parent]," and "My [parent] really cares for me," and a few with negative wording such as "My [parent] is unhappy with the things I do." Possible answers were on a 5-point Likert-type scale ranging from 1 (*Strongly agree*) to 5 (*Strongly disagree*).

A parallel set of questions assessed the adolescent's identification and quality of interactions with his or her sibling in the study (Furman & Buhrmester, 1985). This questionnaire included questions such as "How much do you like the same things," and "How much do you spend free time together." It also included some negatively worded questions such as "How much does your brother or sister tease you?" Possible answers were on a 4-point scale ranging from 1 (*A lot*) to 4 (*Not at all*). As shown in Table 1, alphas for these identification items were .89 or higher.

To add another perspective on supportiveness from family members, we included observer ratings of family interactions. We used a group of five positive behaviors (warmth, listener responsiveness, communication, being prosocial, and relationship quality) and three negative behaviors (hostility, angry coercion, and being antisocial) displayed by each family member toward the target adolescent. Each of the behaviors, positive or negative, was rated by coders as 1 = not characteristic to 5 = mainly characteristic of the family member during the videotaped interactions. The dyad's relationship quality was scored from 1 = negative to 5 = positive. We combined the five positive behaviors by each family member toward the target for an overall score of positive engagement and supportiveness, and we composed a measure of the family member's rejection of the target and lack of support from their three negative behaviors. Alphas for the subscales ranged from .76 to .92, and the overall observer ratings alpha was .89 (see Table 1).

Social support from friends. Two composite variables measured social support from outside of the family as well as related constructs shown to be of assistance in recovering from trauma (e.g., school-related resources). These composite variables were assembled in the same manner as the family variables and the means, standard deviations, and alpha coefficients for each subscale and the total scales are shown in Table 1.

With regard to support from friends, two questionnaires measured perceived supportiveness of the teen's friends: his or her sense of closeness with them and ability to gather support as needed. The first set of positively worded (e.g., "I trust my friends") and negatively worded (e.g., "It is hard for me to make new friends") statements assessed the support available to the adolescent from friends and acquaintances. Respondents answered on a 4-point scale ranging from 1 (*Definitely true*) to 4 (*Definitely false*). The second questionnaire measured direct social support from friends with questions such as "These friends care about me," and included one negatively worded statement: "These friends always criticize me." Respondents answered on a Likert-type 5-point scale ranging from 1 (*Strongly agree*) to 5 (*Strongly disagree*). The alpha (.82) for friend support was quite good (Table 1).

Supportive school environment. The adolescent's feelings about school and the school milieu were evaluated with three variables. Two measures were answered on a 5-point Likert-type scale ranging from 1 (*Strongly agree*) to 5 (*Strongly disagree*). The first (perceived support) was a direct assessment of social support at school and included statements such as "The teachers at my school help students feel more sure of themselves." The second was a measure of the adolescent's sense of attachment to school and school-related activities. It included both positive assertions, such as "In general, I like school a lot," and negative ones, for instance, "I don't feel I really belong at school." These questions indicate the extent to which school is an important and comfortable part of an adolescent's life. To the extent that it is, it should play a role in providing a sense of safety and support during times of extreme stress. The third measure assessed the extent to which the school could, or could not, provide a physically safe environment to the student and with it, a positive milieu for support and learning. The questions deal with school-related problems such as student drug and alcohol use ("How much of a problem has the use of illegal drugs been in your school?") and administrative disorganization ("How much of a problem has poor discipline in the classroom been in your school?"). The respondents could answer on a scale ranging from 1 (*a very serious problem*) to 4 (*no problem at all*). The alpha (.90) for a supportive school environment was more than adequate (Table 1).

Data Analysis

We used logistic regression analyses to investigate associations between traumatic events and psychiatric diagnoses. In logistic regression, the likelihood of a binary outcome (e.g., a diagnosis) given specific circumstances (e.g., experiencing a trauma) is turned into an odds ratio (OR; the odds of receiving the diagnosis under those circumstances divided by the odds under different circumstances). This ratio is transformed through a natural log into a linear function. This analytic strategy provides the best fit to binary outcomes, especially in situations in which the number of cases in outcome categories is very uneven. The results of logistic regression can be reported in two forms: a coefficient, which relates linearly to the logistic function, and the OR, the exponent of the coefficient.

In analyses of psychiatric disorder, the OR is usually reported along with its confidence interval. The confidence interval suggests the reliability of the result and is a function of sample size. Specifically, the confidence interval reflects the number of individuals in the cells under examination (e.g., the number of childhood maltreatment victims who developed drug dependence) and smaller numbers of individuals will result in larger confidence intervals. An OR of 1 means that the odds are equal under the two circumstances (there is no relationship between the predictor and the diagnosis). For that reason, if the confidence interval includes 1, the result is statistically nonsignificant. ORs greater than 1 indicate increased odds of psychiatric disorder for the group being considered in relation to a comparison group and are associated with positive coefficients. ORs below 1 indicate that the odds of the outcome are lower than for the comparison group and are associated with negative coefficients. The degree of risk, or effect size, is measured by the distance of the OR from 1, keeping in mind, however, that ratios above 1 may go up indefinitely, whereas below 1, they are asymptotically bounded by zero.

We followed several conventions in our logistic regression models. First, we always modeled the odds of positive diagnosis versus no diagnosis. Second, we always modeled the risk of being female versus being male. Third, we used the group of individuals who had reported no traumatic events as our comparison group for all analyses except for PTSD; in the UM–CIDI diagnostic interview, there was no assessment for PTSD unless a traumatic event was reported, so in those analyses, we used individuals who experienced traumas after age 15 as our comparison group.

Predictor variables. We used the type of trauma exposure experienced by the respondent to predict psychiatric disorder. The variable, trauma type, has four levels: 0 if the respondent had not experienced a trauma, 1 if the respondent had experienced a trauma after age 15, 2 if the trauma was experienced before age 15 but did not involve maltreatment, and 3 if the respondent was maltreated before age 15. Maltreatment included neglect, physical abuse, and rape or sexual molestation. For purposes of analysis, the levels of this variable were dummy coded into three dichotomous variables with no trauma as the reference category.

Outcome variables. Several available diagnostic outcomes were assessed but excluded from these analyses. Eating disorders, generalized anxiety disorder, and manic and hypomanic episodes were too rare in our sample for analysis (i.e., fewer than 10 cases). Because of omissions such as these, we tested associations with broad categories of diagnoses (e.g., affective disorders). The exception to this was substance use disorders. We reported drug and alcohol dependence rather than the broader inclusive category of "abuse" because dependence is the more serious disorder and because we felt the noteworthy effect sizes warranted attention.

We analyzed adult antisocial behavior rather than antisocial personality disorder. Antisocial personality disorder requires a diagnosis of conduct disorder before age 15 plus a pattern of antisocial behaviors after age 15. The adult antisocial behavior designation, although not a diagnosis, includes all individuals meeting full diagnostic criteria for antisocial personality disorder as well as those who meet all of the criteria except that they lack the earlier diagnosis of childhood conduct disorder. We used this classification rather than conduct disorder, which is only diagnosed before age 15, or the more limited antisocial personality disorder to characterize individuals in our sample with serious behavior problems in adulthood.

Hierarchical regression analysis. The target's sex was entered first in all analyses, and the parents' education and income (family income-to-needs ratio) were entered second. In cases in which education or income had a significant effect on the outcome disorder being investigated, we kept them in the model through remaining steps. In some cases, they had no effect. We did not retain them in the models in these cases. In the next two steps, the predictor variable, trauma type, then the social support variable were entered (social support variables were entered separately). Finally, interactions between each type of social support and the trauma condition were tested.

Missing data. The predictor variable, traumatic experience, came from two diagnostic interviews administered in 1997 and 1999, and the data include all individuals who were present at one or both interviews. We conducted analyses on differences between respondents who were present for one of the two interviews versus present for both. There were no differences between the groups in terms of sex or types or overall number of traumas experienced. They did differ on two diagnoses. Individuals who received the diagnosis of conduct disorder, a childhood diagnosis, were less likely to be present for both 1997 and 1999 interviews, and respondents who were diagnosed with alcohol abuse were more likely to be present for both interviews than those who did not receive this diagnosis. In addition, 35 young people who were evaluated in diagnostic interviews were not present in the 1991 assessments. There were no differences between these individuals and the remaining 489 in terms of sex or trauma experiences, but again, those with conduct disorder diagnoses were more likely to have missed the 1991 assessment. For cases missing 1991 data, we used expectation maximization (EM) imputation to replace the missing values for social support, parental education, and family income-to-needs ratio. Replacement of missing values rather than case deletion is recommended particularly in situations in which the cases with missing values may represent a large proportion of any subpopulation of interest (Schafer & Graham, 2002). In this case, we would risk removing data of individuals who have less common disorders or were victims of trauma. EM imputation has produced less biased estimates of missing values than other common methods of imputation such as mean replacement and also provides more accurate estimates of associations among variables than pairwise or listwise deletion (Schafer & Graham, 2002).

RESULTS

The Prevalence of Traumatic Events

Five hundred twenty-four individuals completed the diagnostic interview in 1997, 1999, or both (285 women, 239 men). Forty-three percent ($n = 228$) reported experiencing one or more traumatic events in their lifetimes (see Table 2). Life-threatening accidents, witnessing serious injury or death, and natural disasters were the most commonly reported traumatic events in this rural community sample, with the shocking revelation of another's trauma following close behind. By 1999, when these young people averaged 23 years of age, none had experienced combat. The Gulf War, from 1990 to 1991, had taken place while they were still under age 16, and the current conflicts in Afghanistan and Iraq had not yet started. The young women in the cohort were far more likely than men to be sexually victimized, whereas the young men were about twice as likely to be involved in serious accidents.

Of the 228 individuals who reported traumatic events, over half of them had experienced only one event ($n = 133$), and a maximum of seven events was reported. Within the entire sample, therefore, 82% ($n = 429$) experienced no events or just one traumatic event (see Table 3). Those individuals who experienced traumatic events of any type after age 15 were likely to have experienced only one event (66%). Children who experienced maltreatment including physical and sexual abuse and neglect were more likely to have experienced two or more traumatic events (75% or 21 children). Figure 1 underscores the importance of type and timing of trauma in the experience of adversity. It shows that respondents who reported a first traumatic event after age 15 averaged about 1.5 traumas, those who reported a serious adversity before age 15 averaged about two traumatic events,

TABLE 2
Table of Prevalence of Traumatic Events by Gender and Age

	Percentage of Sample by									
	Sex				Age at Trauma					
Traumatic Experience	*Males*[a]	*n*	*Females*[b]	*n*	*Under 15*	*n*	*Over 15*	*n*	*Percentage of Total*[c]	*n*
Life-threatening accident (including life-threatening medical emergencies)	20.08	48	9.12	26	6.11	32	8.02	42	14.12	74
Natural disaster	14.64	35	9.47	27	4.77	25	7.06	37	11.83	62
Witnessed death or serious injury	19.25	46	9.12	26	3.82	20	9.92	52	13.74	72
Raped	0.42	1	6.67	19	1.34	7	2.48	13	3.82	20
Sexually molested	—	—	7.02	20	3.24	17	0.57	33	3.82	20
Physically attacked or assaulted	5.86	14	4.56	13	1.72	9	3.44	18	5.15	27
Physically abused as child	0.84	2	3.51	10	2.29	12			2.29	12
Seriously neglected as child	0.84	2	1.40	4	1.15	6			1.15	6
Threatened with a weapon, kidnapped, or held captive	8.79	21	2.46	7	2.29	12	3.05	16	5.34	28
Other serious trauma	3.77	9	2.81	8	1.53	8	1.72	9	3.24	17
Shock at hearing that trauma happened to somebody else	8.79	21	11.58	33	3.24	17	7.06	37	10.31	54
Total reporting at least one event[d]	48.95	117	38.95	111	13.17	69	30.34	159	43.51	228

[a]$n = 239$. [b]$n = 285$. [c]$N = 524$. [d]Some individuals reported more than one event.

TABLE 3
Number of Events Reported by Entire Sample and by Each Subgroup

Number of Traumatic Events Reported	Percentage of Total Sample[a]		Child Maltreatment Victims[b]		Other Early Trauma Victims[c]		Later Trauma Victims[d]	
	%	*n*	%	*n*	%	*n*	%	*n*
0	56.49	296						
1	25.38	133	25.00	7	51.22	21	66.04	105
2	8.97	47	14.29	4	14.63	6	23.27	37
3	5.53	29	21.43	6	21.95	9	8.81	14
4	1.91	10	10.71	3	9.76	4	1.89	3
5	0.76	4	14.29	4	—	—	—	—
6	0.57	3	10.71	3	—	—	—	—
7	0.38	2	3.57	1	2.44	1	—	—
Total	406		90		83		233	

[a]$N = 524$. [b]$n = 28$. [c]$n = 41$. [d]$n = 159$.

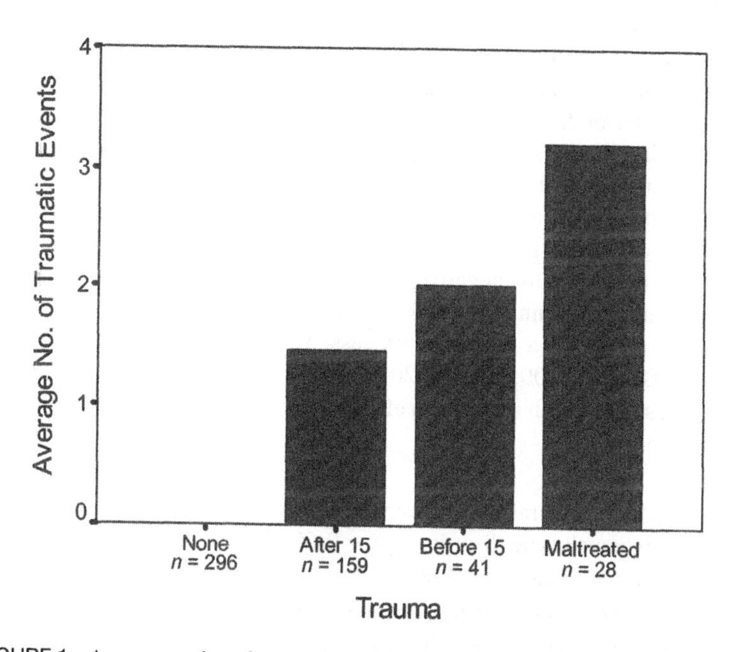

FIGURE 1 Average number of traumatic events experienced through age 23 by trauma condition.

and the maltreated children averaged more than three serious traumas by age 23. These differential experiences are an important consideration when examining the impact of trauma in psychiatric disorders.

Traumatic Events, Social Support, and Psychiatric Disorder

Earlier research (Kessler et al., 1994) indicated that women are at greater risk than men for affective and anxiety disorders and that men are at greater risk for behavior and substance use disorders. Because of these gender differences in psychopathology, we controlled for sex in all analyses. We coded female as 1 and male as 0, so ORs for sex that are greater than 1 indicated increased risk of females relative to males, whereas ORs less than 1 indicated the opposite. We also evaluated Trauma × Gender interactions in predicting psychiatric disorder. The number of statistically significant interactions did not exceed chance expectations at the .05 level; therefore, we do not report them here and we present all analysis for the combined sample. In addition, family characteristics suggestive of low socioeconomic status have also been associated with psychiatric disorder in children (Sameroff, 2000). We found that family income-to-needs ratio and the average of mother's and father's education had main effects for a few classes of disorders. Specifically, one or the other predicted behavior disorders and the likelihood overall of being diagnosed with any disorder. For these analyses, we retained them in the models as control variables.

Researchers interested in processes of resilience have proposed that evidence for resilience can occur in at least two forms. Specifically, resilience-promoting mechanisms can reduce the negative impacts of life's adversities either through a significant statistical main effect (a compensatory process) or through a significant interaction effect (a buffering process; Conger & Conger, 2002; Luthar, Cicchetti, & Becker, 2000; Masten, 2001). We examined both of these possibilities, and in cases in which interaction effects were found, we report them in the tables that follow.

Any disorder. Overall, males were at somewhat greater risk than females of developing any psychiatric disorder (OR = 0.61, $p < .01$; see Table 4). A significant OR for parents' education (OR = .88, $p < .05$) suggests that children whose parents are better educated are at lower risk of disorder, although this OR must be interpreted with caution because the change in fit when this variable, along with family income-to-needs ratio, was entered in the model was not statistically significant. Nonetheless, the parents' education and income variables were retained in the model. Table 4 shows how exposure to trauma increases the risk of developing any disorder, after controlling for sex and parental demographic character-

TABLE 4
Conditional Risks of Psychiatric Disorder: Any Disorder

	OR	*OR CI*	*df*	χ^2	Δdf	$\Delta\chi^2$
Step 1: Control variables						
Sex (1 = female)	0.61**	0.43–0.88	1	688.66	1	7.16**
Step 2: Control variables			3	683.31	2	5.35
Parents' income	0.99	0.91–1.07				
Parents' education (average)	0.88*	0.77–1.00				
Step 2: Trauma type			6	662.26	3	26.40**
Childhood maltreatment	5.64**	1.89–16.89				
Other early trauma	1.23	0.62–2.44				
Later trauma	2.02**	1.33–3.07				
Step 3: Social support						
Family	0.74**	0.60–0.90	7	652.93	1	9.33**
Friends	0.91	0.74–1.11	7	661.37	1	0.89
School	0.77**	0.63–0.94	7	655.22	1	7.04**

Note. $N = 524$. The no-trauma group is the comparison group in all analyses. OR = odds ratio; OR CI = the 95% confidence interval around the ORs. No interactions were significant.
$^{*}p < .05$. $^{**}p < .01$.

istics. The risk for those who experienced trauma after age 15 was double that of individuals who had experienced no trauma (OR = 2.02, $p < .01$; see Table 4), and it is particularly notable that the risk of disorder for the maltreated children was more than five times that of the no-trauma group (OR = 5.64, $p < .01$; Table 4).

Social support from family and a supportive school atmosphere provided overall protection from disorder (ORs = 0.74 and 0.77, respectively, both $ps < .01$). In this general diagnostic group, there were no statistical interactions with social support. That is, social support was helpful to everybody whether or not they experienced trauma, and if they did experience trauma, social support had a compensatory main effect in reducing risk for disorder regardless of the type of traumatic experiences they had or the age at which they experienced them.

Affective disorders. Females demonstrated more than three times the risk for developing affective disorders as males (OR = 3.30, $p < .01$; Table 5). After statistically controlling for sex, however, childhood maltreatment still proved a strong predictor of affective disorders as shown both by an almost sevenfold increase in risk for those who were maltreated (OR = 6.84, $p < .01$) and by a substantial improvement in fit brought about by the addition of trauma type to the model, $\Delta\chi^2(3, N = 524) = 20.36$, $p < .01$.

Some forms of social support proved protective in the case of affective disorders. Although a significant main effect suggested that all young people in this study benefited from social support, especially in their school settings, the significant interaction ORs, which were less than 1, indicated a greater reduction in risk

TABLE 5
Conditional Risks of Psychiatric Disorder: Any Affective Disorder

	OR	*OR CI*	*df*	χ^2	Δdf	$\Delta\chi^2$
Step 1: Control variables						
Sex (1 = female)	3.30**	1.93–5.63	1	445.98	1	21.89**
Step 2: Trauma type			4	425.62	3	20.36**
Childhood maltreatment	6.84**	2.97–15.75				
Other early trauma	1.40	0.54–3.63				
Later trauma	1.20	0.69–2.12				
Step 3: Social support						
Family	0.91	0.71–1.17	5	425.11	1	0.51
Friends	0.79	0.62–1.02	5	422.49	1	3.13
School	0.75*	0.58–0.96	5	420.17	1	5.45*
Step 4: Interactions						
Friends' support by trauma type			8	415.47	3	7.02
Maltreatment	0.87	0.44–1.73				
Other early trauma	0.23*	0.06–0.80				
Later trauma	0.65	0.36–1.18				
School support by trauma type			8	411.06	3	9.11*
Maltreatment	0.39*	0.16–0.95				
Other early trauma	0.41	0.15–1.10				
Later trauma	0.52*	0.29–0.93				

Note. $N = 524$. The no-trauma group is the comparison group in all analyses. OR = odds ratio; OR CI = the 95% confidence interval around the OR. Parents' income and education had no significant main effects so were omitted from further analyses. Only significant interactions are shown.
*$p < .05$. **$p < .01$.

for those young people who suffered traumatic events than for youth who did not. An example may be seen in Figure 2 in which social support from school, measured in standard deviation units with a mean of zero, is shown on the x-axis, and odds of any affective disorder is shown on the y-axis. Here, a significant interaction with maltreated children (OR = 0.39, $p < .05$) and one of lesser magnitude with later trauma victims (OR = 0.52, $p < .05$) are shown by the differences in slopes for the different groups. All trauma victims were at greater risk of developing affective disorders than the participants who did not experience trauma when social support levels were very low. As everybody's social support increased, however, risk levels dropped for trauma victims. Because risk levels were already extremely low for the no-trauma group, they did not change. Maltreatment victims started at strikingly greater disadvantage than non-maltreated victims; thus their decline in risk was more dramatic (the slope is steeper) as their school social support increased. The individuals who experienced traumatic events after age 15 had a significantly different slope from those who experienced no traumas, whereas the younger victims who were not maltreated did not differ significantly (OR = 0.41, $p < .08$) despite the fact that the slope of the younger victims was

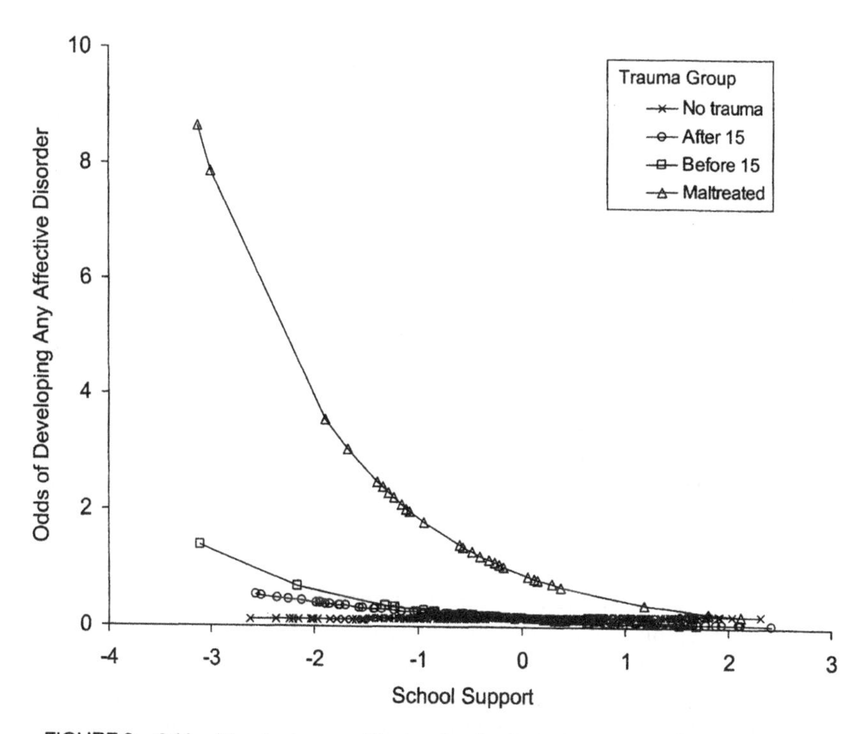

FIGURE 2 Odds of developing any affective disorder for each trauma condition subgroup depending on level of social support from school in ninth grade. Social support is measured in standard deviation units with a mean of zero.

steeper and their initial risk greater. This was probably the result of the smaller number of early trauma victims ($n = 41$) compared with later victims ($n = 159$). It is also interesting to note that the risk of affective disorders for the maltreated children not only started high, but remained elevated until their social support levels were more than one standard deviation above the sample mean.

In addition to the interaction with school-related support, we showed a significant interaction OR (OR = 0.23, $p < .05$) with support from friends for the nonmaltreated, early-trauma victims. The improvement to model fit with the addition of this interaction, however, did not quite reach significance, $\Delta\chi^2(3, N = 524) = 7.02$, $p = .07$, so this finding must be viewed with some caution.

Anxiety disorders. We examined the changes in the ORs for anxiety disorders as a whole and also for PTSD alone. Because PTSD is by definition associated with the experience of trauma, individuals who reported no traumatic events were not asked questions that could lead to the diagnosis in this interview. Therefore, the overall sample was smaller in these analyses ($n = 228$). The comparison group for the PTSD ORs was the group that reported traumatic

TABLE 6
Conditional Risks of Psychiatric Disorder: Posttraumatic Stress Disorder

	OR	*OR CI*	*df*	χ^2	Δdf	$\Delta\chi^2$
Step 1: Control variables						
Sex (1 = female)	6.59**	2.18–19.89	1	142.55	1	15.13**
Step 2: Trauma type			3	136.49	2	6.06*
Childhood maltreatment	3.80*	1.33–10.30				
Other early trauma	1.64	0.48–5.61				
Step 3: Social support						
Family	1.09	0.70–1.71	4	136.34	1	0.15
Friends	0.83	0.52–1.31	4	135.83	1	0.66
School	0.70	0.46–1.06	4	133.61	1	2.88

Note. $N = 228$. The group that experienced traumas after age 15 were the comparison group. OR = odds ratio; OR CI = the 95% confidence interval around the OR. Parents' income and education had no significant main effects so were omitted from further analyses. No interactions were significant.

$^*p < .05$. $^{**}p < .01$.

events after age 15. Consistent with the literature (Brewin et al., 2000), females were at considerably greater risk of developing PTSD than are males (OR = 6.59, $p < .01$; Table 6) when both have experienced traumatic events. Compared with adult trauma victims, childhood maltreatment victims were at nearly four times the risk of developing PTSD (OR = 3.80, $p < .05$). We found no significant protective effects of social support within our sample (Table 6) when it came to PTSD.

The story is slightly different with the overall category of anxiety disorders. Here again, females were at greater risk, and both maltreatment victims and individuals who had experienced more recent traumatic events showed a significantly elevated risk of developing any anxiety disorder (see Table 7). Nonmaltreated childhood trauma victims also showed an increased risk of anxiety disorder, although the OR was only marginally significant (OR = 2.11, $p = .05$). Moreover, increased school support showed protective effects for the development of anxiety disorders for all individuals irrespective of trauma exposure.

Adult antisocial behavior. Males were at greater risk of adult antisocial behavior (OR = 0.50, $p < .05$) as were children from families at lower income levels (OR = 0.67, $p < .01$; see Table 8). Here again, victims of childhood maltreatment were at greatest risk of demonstrating serious antisocial behaviors in adulthood (OR = 7.40, $p < .01$). Traumatic events experienced after age 15 were also associated with increased risk for antisocial behavior. Consistent with expectations, both family and school support demonstrated compensatory effects resulting in decreased risk for all of these young adults. There was no evidence, however, for buffering effects for adult antisocial behavior.

TABLE 7
Conditional Risks of Psychiatric Disorder: Any Anxiety Disorder

	OR	*OR CI*	*df*	χ^2	Δdf	$\Delta\chi^2$
Step 1: Control variables						
Sex (1 = female)	1.81**	1.21–2.71	1	593.67	1	8.48**
Step 2: Trauma type			4	558.13	3	35.54**
Childhood maltreatment	9.69**	4.03–23.30				
Other early trauma	2.11	1.00–4.47				
Later trauma	2.19**	1.39–3.44				
Step 3: Social support						
Family	0.92	0.75–1.14	5	557.58	1	0.55
Friends	0.82	0.66–1.02	5	554.95	1	3.18
School	0.74*	0.60–0.91	5	550.00	1	8.13*

Note. $N = 524$. The no-trauma group is the comparison group in all analyses. OR = odds ratio; OR CI = the 95% confidence interval around the OR. Parents' income and education had no significant main effects so were omitted from further analyses. No interactions were significant.
$^*p < .05$. $^{**}p < .01$.

TABLE 8
Conditional Risks of Psychiatric Disorder: Adult Antisocial Behavior

	OR	*OR CI*	*df*	χ^2	Δdf	$\Delta\chi^2$
Step 1: Control variables						
Sex (1 = female)	0.50*	0.29–0.89	1	350.48	1	5.76*
Step 2: Control variables			3	336.56	2	13.92**
Parents' income	0.67**	0.52–0.86				
Parents' education (average)	1.06	0.89–1.25				
Step 3: Trauma type			6	319.37	3	17.19**
Childhood maltreatment	7.40**	2.75–19.94				
Other early trauma	1.33	0.42–4.25				
Later trauma	2.33*	1.22–4.46				
Step 4: Social support						
Family	0.70*	0.52–0.96	7	314.46	1	4.91*
Friends	1.00	0.73–1.37	7	319.37	1	0.00
School	0.58**	0.43–0.79	7	306.53	1	12.84**

Note. $N = 524$. The no-trauma group is the comparison group in all analyses. OR = odds ratio; OR CI = the 95% confidence interval around the OR. No interactions were significant.
$^*p < .05$. $^{**}p < .01$.

Substance use disorders. Substance use disorders may be classified into dependence and abuse disorders in which dependence is the more severe of the two. For that reason, we focused our analyses on alcohol dependence and drug dependence. In both cases, males were at considerably greater risk of developing dependence than females (OR = 0.26, $p < .01$, alcohol dependence; OR = 0.23, $p < .01$, drug dependence; see Table 9), and maltreatment victims were again at the greatest risk of disorder (OR = 3.84, $p < .01$, alcohol dependence; OR = 18.54, $p < .01$, drug dependence; Table 9). In both cases, school support reduced risk for substance dependence in a compensatory process. Of particular note were the large magnitudes of the increases in risk of substance dependence other than alcohol for individuals who suffered traumas early in life, especially those who suffered maltreatment. This increase in risk was reflected in both large ORs and a substantial improvement in fit when the trauma-group variables were added to the model, $\Delta\chi^2(3, N = 480) = 16.23, p < .01$ (Table 9). Finally, later trauma also predicted a diagnosis of alcohol dependence.

DISCUSSION

Overall, traumatic experiences were not rare in our sample of young adults, but neither were they common occurrences. The frequency of trauma in our sample

TABLE 9
Conditional Risks of Psychiatric Disorder: Substance Dependence Disorders

	OR	*OR CI*	*df*	χ^2	Δdf	$\Delta\chi^2$
Alcohol Dependence						
Step 1: Control variables						
Sex (1 = female)	0.26**	0.16–0.42	1	479.99	1	33.72**
Step 2: Trauma type			4	470.42	3	9.57*
Childhood maltreatment	3.84**	1.50–9.82				
Other early trauma	1.19	0.52–2.74				
Later trauma	1.68*	1.02–2.77				
Step 3: Social support						
Family	0.82	0.64–1.05	5	467.96	1	2.46
Friends	0.89	0.71–1.12	5	469.46	1	0.96
School	0.72**	0.57–0.91	5	462.94	1	7.48**
Drug dependence, not alcohol						
Step 1: Control variables						
Sex (1 = female)	0.23**	0.08–0.63	1	168.75	1	9.87**
Step 2: Trauma type			4	152.52	3	16.23**
Childhood maltreatment	18.54**	4.01–85.72				
Other early trauma	6.18**	1.66–23.04				
Later trauma	2.56	0.81–8.08				
Step 3: Social support						
Family	0.87	0.53–1.45	5	152.26	1	0.26
Friends	0.82	0.54–1.25	5	151.71	1	0.81
School	0.64*	0.41–0.99	5	148.44	1	4.08*

Note. $N = 524$ for alcohol dependence; $N = 480$ for drug dependence. The no-trauma group is the comparison group in all analyses. OR = odds ratio; OR CI = the 95% confidence interval around the OR. Parents' income and education had no significant main effects so were omitted from further analyses. No interactions were significant.
*$p < .05$. **$p < .01$.

was the same as that found by Giaconia et al. (1995) for their sample of 18-year-olds; they also reported a prevalence rate of 43%. With the exception of combat, the distribution of types of trauma was very much like that found in the National Comorbidity Study (Kessler et al., 1995), which also used the UM–CIDI as a diagnostic tool.

The majority of these young people reported no traumatic experiences or just one event. However, most of those who suffered maltreatment as children (including physical and sexual abuse and neglect) were more likely to suffer multiple traumas during their lives. Only 25% of those in the maltreatment group reported just a single life trauma. The remainder reported from two to seven traumas before age 23. In a number of cases, children suffered more than one form of abuse. In addition, many of these victims of maltreatment suffered from psychiatric disorders that put them into situations in which they were at elevated risk of further traumas (e.g., behavior and substance disorders), or put them at risk of self-harm (e.g., affective disorders; Brown et al., 1999). Finally, these young people remained at risk of being revictimized later on in their adult lives (Irwin, 1999). These findings suggest that maltreated children represent an especially vulnerable population in terms of increased risk both for additional traumatic experiences and for psychiatric disorder.

In concordance with the findings from other research on trauma and disorder (e.g., Breslau, Davis, Andreski, & Peterson, 1991; Cohen, Brown, & Smailes, 2001; Fergusson et al., 1996; Fergusson & Lynskey, 1997; Kessler et al., 1995), our data showed that suffering one or more traumatic events increased the likelihood that an individual would also suffer a psychiatric disorder of some type. The bulk of the risk was carried by those maltreated in childhood. They were more likely than individuals who had experienced no trauma to develop every type of disorder we evaluated and were 3.8 times more likely than those who experienced later traumas to develop PTSD as a direct result of their traumatic experiences. Although the clear delineation of temporal order between predictors and outcomes was difficult with retrospective data, our results were consistent with other research that has shown that children who experienced traumatic events, especially maltreatment, were at greater than average risk for affective, anxiety, and behavior disorders over time (Cohen et al., 2001; Dodge et al., 1997).

The results also demonstrate resilience-promoting main effects for social support across most diagnostic categories. As has been reported in most earlier research, these compensatory processes were more common than buffering effects (Luthar et al., 2000; Masten, 2001). Family proved to be an important source of support, reducing the risk for behavior problems in early adulthood. It is also worth noting that family support was broadly protective when we predicted risk for any type of disorder compared to no evidence of disorder. Adolescents who felt a strong sense of support from school personnel, were comfortable with academic tasks, and had a sense of safety and security at school had advantages across the board when it came to mental health outcomes. They were less likely to be diagnosed with behavior, affective, and anxiety disorders as well as less likely to become dependent on drugs or alcohol. In addition, like family support, supportive schools were broadly protective against disorders in general. Buffering effects for social support were found only in relation to affective disorders. Significant interactions between support and trauma were found for both friend and school support. However, it was support from school that played a particularly important role in reducing risk of an affective disorder for the highly vulnerable maltreatment group.

The importance of school, relative to family, for these young people may have been a function of the age at which these assessments were made. In the ninth grade, when the assessments of social support were made, an adolescent is focused on relationships outside of the family and involved in the normative processes of individuation from parents (Grotevant, 1998; Steinberg & Morris, 2000). Life in general, and especially social life, revolves largely around school. In addition, often during adolescence family life is characterized by some degree of conflict as the child separates from home life. This should be especially true for maltreated children. This could decrease the impact of social support at home, especially if the support is offered by parents.

Interestingly, supportive friendships were not associated as a main effect with lower odds of disorder. This result is somewhat surprising with respect to affective and anxiety disorders and may reflect a failure of our measures to capture the relevant aspects of social support from friends. However, a significant interaction suggested that supportive friends may have been uniquely helpful to child trauma victims who were not maltreated when it came to affective disorders. This interaction, however, must be interpreted with some caution, as the associated improvement in fit, signified by the chi-square statistic, was marginal but not significant. It is not surprising that support from friends did not decrease risk for behavior and substance use disorders. This result is consistent with theory that links association with deviant peers in adolescence rather than peer rejection with externalizing disorders (e.g., Rudolph et al., 2000).

In interpreting these compensatory and buffering effects of social support, the possibility must be kept in mind that characteristics of the individuals who encourage warm, supportive friendships and family relationships may coincide with lower risk for affective and other disorders (Ensel & Lin, 1991; Luthar et al., 2000). Nonetheless, whatever qualities of the individual and environment were reflected in these variables, these analyses demonstrate that social support in various domains appears to promote resilience to trauma through both compensatory and buffering processes.

As always, specific limitations of the study must be noted. First, these analyses were constrained by the small size of the subgroups that were at greatest risk for disorder and the relative infrequency of specific psychiatric diagnoses and traumatic events in a nonclinical sample. For instance, variations in risk for disorder have been found to exist between different types of childhood maltreatment (e.g., Cohen et al., 2001; Kilpatrick et al., 2003) as well as between other types of trauma. For example, rape victims are at special risk for PTSD (Kessler et al., 1995), but we did not have enough statistical power to look for these differences in our sample. Second, our sample was rural and of European heritage; thus, these results call for replication with other ethnic groups and samples from more varied geographic locations.

Third, although our social support data were gathered when the target youth were in ninth grade, the trauma and diagnostic assessments were made when they were in their early 20s and so suffer from a weakness common to all retrospective data; they depend on the accuracy of the respondent's memory. Memory for symptoms may be subject to mood-dependent effects and the exact timing of events, and symptom onsets may not be accurate. More important, some events may be forgotten altogether no matter how great their impact at the time or may be too uncomfortable to discuss no matter how long ago they occurred. Furthermore, the lack of precision in the dating of childhood disorders and maltreatment did not allow us to test possible causal links. For instance, Stouthamer-Loeber, Loeber, Homish, and Wei (2001) found that although most often children's development of disruptive behavior followed maltreatment, occasionally maltreatment was preceded by children's behavior problems. Additionally, Kim, Conger, Elder, and Lorenz (2003) showed that the relation between stressful life events and internalizing and externalizing symptoms is reciprocal in adolescents. However, even with 100% prospective data, it is seldom possible to tease apart causes and effects when studying complex, interactive processes. Of perhaps greater importance than finding causes is understanding the processes themselves because it is by intervening in processes that we may affect outcomes. Our hope is that this article can make a contribution to achieving these goals.

ACKNOWLEDGMENTS

During the past several years, support for this research has come from multiple sources, including the National Institute of Mental Health (MH00567, MH19734, MH43270, MH48165, and MH51361), the National Institute on Drug Abuse (DA05374), the Bureau of Maternal and Child Health (MCJ-109572), the McArthur Foundation Research Network on Successful Adolescent Development Among Youth in High-Risk Settings, and the Iowa Agriculture and Home Economics Experiment Station (Project No. 3320).

REFERENCES

American Psychiatric Association. (1987). *Diagnostic and statistical manual of mental disorders* (3rd ed., rev.). Washington, DC: Author.

American Psychiatric Association. (2000). *Diagnostic and statistical manual of mental disorders* (4th ed., text rev.). Washington, DC: Author.

Benotsch, E. G., Brailey, K., Vasterling, J. J., Uddo, M., Constans, J. I., & Sutker, P. B. (2000). War zone stress, personal and environmental resources, and PTSD symptoms in Gulf War veterans: A longitudinal perspective. *Journal of Abnormal Psychology, 109*, 205–213.

Bolton, D., O'Ryan, D., Udwin, O., Boyle, S., & Yule, W. (2000). The long-term psychological effects of a disaster experienced in adolescence: II. General psychopathology. *Journal of Child Psychology and Psychiatry and Allied Disciplines, 41*, 513–523.

Breslau, N., Davis, G. C., Andreski, P., & Peterson, E. (1991). Traumatic events and posttraumatic stress disorder in an urban population of young adults. *Archives of General Psychiatry, 48*, 216–222.

Breslau, N., Kessler, R. C., Chilcoat, H. D., Schultz, L. R., Davis, G. C., & Andreski, P. (1998). Trauma and posttraumatic stress disorder in the community: The 1996 Detroit Area Survey of Trauma. *Archives of General Psychiatry, 55*, 626–632.

Brewin, C. R., Andrews, B., & Valentine, J. D. (2000). Meta-analysis of risk factors for posttraumatic stress disorder in trauma-exposed adults. *Journal of Consulting and Clinical Psychology, 68*, 748–766.

Brown, J., Cohen, P., Johnson, J. G., & Salzinger, S. (1998). A longitudinal analysis of risk factors for child maltreatment: Findings of a 17-year prospective study of officially recorded and self-reported child abuse and neglect. *Child Abuse & Neglect, 22*, 1065–1078.

Brown, J., Cohen, P., Johnson, J. G., & Smailes, E. M. (1999). Childhood abuse and neglect: Specificity and effects on adolescent and young adult depression and suicidality. *Journal of the American Academy of Child & Adolescent Psychiatry, 38*, 1490–1496.

Cicchetti, D., Toth, S. L., & Lynch, M. (1997). Child maltreatment as an illustration of the effects of war on development. In S. L. Toth (Ed.), *Developmental perspectives on trauma: Theory, research, and intervention. Rochester symposium on developmental psychology* (Vol. 8, pp. 227–262). Rochester, NY: University of Rochester Press.

Cohen, P., Brown, J., & Smailes, E. (2001). Child abuse and neglect and the development of mental disorders in the general population. *Development & Psychopathology, 13*, 981–999.

Conger, R. D., & Conger, K. J. (2002). Resilience in Midwestern families: Selected findings from the first decade of a prospective, longitudinal study. *Journal of Marriage and Family, 64*, 361–373.

Conger, R. D., & Elder, G. H. (1994). *Families in troubled times: Adapting to change in rural America.* Hawthorne, NY: Aldine de Gruyter.

Costello, E. J., Erkanli, A., Fairbank, J. A., & Angold, A. (2002). The prevalence of potentially traumatic events in childhood and adolescence. *Journal of Traumatic Stress, 15*, 99–112.

Dodge, K. A., Pettit, G. S., & Bates, J. E. (1997). How the experience of early physical abuse leads children to become chronically aggressive. In S. L. Toth (Ed.), *Developmental perspectives on trauma: Theory, research, and intervention. Rochester symposium on developmental psychology* (Vol. 8, pp. 263–288). Rochester, NY: University of Rochester Press.

Duraković-Belko, E., Kulenović, A., & Đapić, R. (2003). Determinants of posttraumatic adjustment in adolescents from Sarajevo who experienced war. *Journal of Clinical Psychology, 59*, 27–40.

Edwards, V. J., Holden, G. W., Felitti, V. J., & Anda, R. F. (2003). Relationship between multiple forms of childhood maltreatment and adult mental health in community respondents: Results from the Adverse Childhood Experiences study. *American Journal of Psychiatry, 160*, 1453–1460.

Egeland, B. (1997). Mediators of the effects of child maltreatment on developmental adaptation in adolescence. In S. L. Toth (Ed.), *Developmental perspectives on trauma: Theory, research, and inter-*

vention. Rochester symposium on developmental psychology (Vol. 8, pp. 403–434). Rochester, NY: University of Rochester Press.

Elbedour, S., ten Bensel, R., & Bastien, D. T. (1993). Ecological integrated model of children in war: Individual and social psychology. *Child Abuse & Neglect, 17*, 805–819.

Elder, G. H., Jr., & Conger, R. D. (2000). *Children of the land: Adversity and success in rural America*. Chicago: University of Chicago Press.

Elklit, A. (2002). Victimization and PTSD in a Danish national youth probability sample. *Journal of the American Academy of Child and Adolescent Psychiatry, 41*, 174–181.

Ensel, W. M., & Lin, N. (1991). The life stress paradigm and psychological distress. *Journal of Health & Social Behavior, 32*, 321–341.

Farley, M., & Patsalides, B. M. (2001). Physical symptoms, posttraumatic stress disorder, and healthcare utilization of women with and without childhood physical and sexual abuse. *Psychological Reports, 89*, 595–606.

Fergusson, D. M., Horwood, L. J., & Lynskey, M. T. (1996). Childhood sexual abuse and psychiatric disorder in young adulthood: II. Psychiatric outcomes of childhood sexual abuse. *Journal of the American Academy of Child & Adolescent Psychiatry, 35*, 1365–1374.

Fergusson, D. M., & Lynskey, M. T. (1997). Physical punishment/maltreatment during childhood and adjustment in young adulthood. *Child Abuse & Neglect, 21*, 617–630.

Finkelhor, D., & Kendall-Tackett, K. (1997). A developmental perspective on the childhood impact of crime, abuse, and violent victimization. In S. L. Toth (Ed.), *Developmental perspectives on trauma: Theory, research, and intervention. Rochester symposium on developmental psychology* (Vol. 8, pp. 1–32). Rochester, NY: University of Rochester Press.

Fletcher, K. E. (2003). Childhood posttraumatic stress disorder. In R. A. Barkley (Ed.), *Child psychopathology* (2nd ed., pp. 330–371). New York: Guilford.

Furman, W., & Buhrmester, D. (1985). Children's perceptions of the qualities of sibling relationships. *Child Development, 56*, 448–461.

Giaconia, R. M., Reinherz, H. Z., Silverman, A. B., Pakiz, B., Frost, A. K., & Cohen, E. (1995). Traumas and posttraumatic stress disorder in a community population of older adolescents. *Journal of the American Academy of Child and Adolescent Psychiatry, 34*, 1369–1380.

Ginzburg, K., Solomon, Z., Dekel, R., & Neria, Y. (2003). Battlefield functioning and chronic PTSD: Associations with perceived self-efficacy and causal attribution. *Personality and Individual Differences, 34*, 463–476.

Goenjian, A. K., Pynoos, R. S., Steinberg, A. M., Najarian, L. M., Asarnow, J. R., Karayan, I., et al. (1995). Psychiatric comorbidity in children after the 1988 earthquake in Armenia. *Journal of the American Academy of Child and Adolescent Psychiatry, 34*, 1174–1184.

Green, B. L., Korol, M., Grace, M. C., Vary, M. G., Leonard, A. C., Gleser, G. C., et al. (1991). Children and disaster: Age, gender, and parental effects on PTSD symptoms. *Journal of the American Academy of Child and Adolescent Psychiatry, 30*, 945–951.

Grotevant, H. D. (1998). Adolescent development in family contexts. In N. Eisenberg (Ed.), *Social, emotional, and personality development* (5th ed., Vol. 3, pp. 1097–1149). New York: Wiley.

Heim, C., & Nemeroff, C. B. (2001). The role of childhood trauma in the neurobiology of mood and anxiety disorders: Preclinical and clinical studies. *Biological Psychiatry, 49*, 1023–1039.

Irwin, H. J. (1999). Violent and nonviolent revictimization of women abused in childhood. *Journal of Interpersonal Violence, 14*, 1095–1110.

Kessler, R. C. (2000). Posttraumatic stress disorder: The burden to the individual and to society. *Journal of Clinical Psychiatry, 61*(Suppl. 5), 4–14.

Kessler, R. C., McGonagle, K. A., Zhao, S., Nelson, C. B., Hughes, M., Eshleman, S., et al. (1994). Lifetime and 12-month prevalence of *DSM–III–R* psychiatric disorders in the United States: Results from the National Comorbidity Study. *Archives of General Psychiatry, 51*, 8–19.

Kessler, R. C., Sonnega, A., Bromet, E., Hughes, M., & Nelson, C. B. (1995). Posttraumatic stress disorder in the National Comorbidity Survey. *Archives of General Psychiatry, 52*, 1048–1060.

Kessler, R. C., Sonnega, A., Bromet, E., Hughes, M., Nelson, C. B., & Breslau, N. (1999). Epidemiological risk factors for trauma and PTSD. In R. Yehuda (Ed.), *Risk factors for posttraumatic stress disorder* (pp. 23–59). Washington, DC: American Psychiatric Association.

Kilpatrick, D. G., Ruggiero, K. J., Acierno, R., Saunders, B. E., Resnick, H. S., & Best, C. L. (2003). Violence and risk of PTSD, major depression, substance abuse/dependence, and comorbidity: Results from the National Survey of Adolescents. *Journal of Consulting & Clinical Psychology, 71*, 692–700.

Kim, K. J., Conger, R. D., Elder, G. H., Jr., & Lorenz, F. O. (2003). Reciprocal influences between stressful life events and adolescent internalizing and externalizing problems. *Child Development, 74,* 127–143.

King, D. W., King, L. A., Foy, D. W., & Gudanowski, D. M. (1996). Prewar factors in combat-related posttraumatic stress disorder: Structural equation modeling with a national sample of female and male Vietnam veterans. *Journal of Consulting and Clinical Psychology, 64*, 520–531.

Korol, M., Kramer, T. L., Grace, M. C., & Green, B. L. (2002). Dam break: Long-term follow-up of children exposed to the Buffalo Creek disaster. In W. K. Silverman (Ed.), *Helping children cope with disasters and terrorism* (pp. 241–257). Washington, DC: American Psychological Association.

Lipschitz, D. S., Rasmusson, A. M., & Southwick, S. M. (1998). Childhood posttraumatic stress disorder: A review of neurobiologic sequelae. *Psychiatric Annals, 28*, 452–457.

Luthar, S. S., Cicchetti, D., & Becker, B. (2000). The construct of resilience: A critical evaluation and guidelines for future work. *Child Development, 71*, 543–562.

Macklin, M. L., Metzger, L. J., Litz, B. T., McNally, R. J., Lasko, N. B., Orr, S. P., et al. (1998). Lower precombat intelligence is a risk factor for posttraumatic stress disorder. *Journal of Consulting and Clinical Psychology, 66*, 323–326.

Masten, A. S. (2001). Ordinary magic: Resilience processes in development. *American Psychologist, 56*, 227–238.

Melby, J. N., & Conger, R. D. (2001). The Iowa Family Interaction Rating Scales: Instrument summary. In K. M. Lindahl (Ed.), *Family observational coding systems: Resources for systemic research* (pp. 33–58). Mahwah, NJ: Lawrence Erlbaum Associates, Inc.

Nishith, P., Mechanic, M. B., & Resick, P. A. (2000). Prior interpersonal trauma: The contribution to current PTSD symptoms in female rape victims. *Journal of Abnormal Psychology, 109*, 20–25.

Norris, F. H. (1992). Epidemiology of trauma: Frequency and impact of different potentially traumatic events on different demographic groups. *Journal of Consulting and Clinical Psychology, 60*, 409–418.

Osofsky, J. D., & Scheeringa, M. S. (1997). Community and domestic violence exposure: Effects on development and psychopathology. In S. L. Toth (Ed.), *Developmental perspectives on trauma: Theory, research, and intervention. Rochester symposium on developmental psychology* (Vol. 8, pp. 155–180). Rochester, NY: University of Rochester Press.

Pine, D. S., & Cohen, J. A. (2002). Trauma in children and adolescents: Risk and treatment of psychiatric sequelae. *Biological Psychiatry, 51*(7), 519–531.

Pynoos, R. S., Steinberg, A. M., & Piacentini, J. C. (1999). A developmental psychopathology model of childhood traumatic stress and intersection with anxiety disorders. *Biological Psychiatry, 46*, 1542–1554.

Pynoos, R. S., Steinberg, A. M., & Wraith, R. (1995). A developmental model of childhood traumatic stress. In D. Cicchetti & D. J. Cohen (Eds.), *Developmental psychopathology: Vol. 2. Risk, disorder, and adaptation* (pp. 72–95). Oxford, England: Wiley.

Resnick, H. S., Kilpatrick, D. G., Dansky, B. S., Saunders, B. E., & Best, C. L. (1993). Prevalence of civilian trauma and posttraumatic stress disorder in a representative national sample of women. *Journal of Consulting and Clinical Psychology, 61*, 984–991.

Rossman, B. R., Bingham, R. D., & Emde, R. N. (1997). Symptomatology and adaptive functioning for children exposed to normative stressors, dog attack, and parental violence. *Journal of the American Academy of Child and Adolescent Psychiatry, 36*, 1089–1097.

Royse, D., Rompf, E. L., & Dhooper, S. S. (1991). Childhood trauma and adult life satisfaction in a random adult sample. *Psychological Reports, 69*, 1227–1231.

Rudolph, K. D., Hammen, C., Burge, D., Lindberg, N., Herzberg, D., & Daley, S. E. (2000). Toward an interpersonal life-stress model of depression: The developmental context of stress generation. *Development & Psychopathology, 12*, 215–234.

Rutter, M., & Maughan, B. (1997). Psychosocial adversities in childhood and adult psychopathology. *Journal of Personality Disorders, 11*, 4–18.

Saigh, P. A., Yasik, A. E., Sack, W. H., & Koplewicz, H. S. (1999). Child-adolescent posttraumatic stress disorder: Prevalence, risk factors, and comorbidity. In J. D. Bremner (Ed.), *Posttraumatic stress disorder: A comprehensive text* (pp. 18–43). Needham Heights, MA: Allyn & Bacon.

Sameroff, A. J. (2000). Developmental systems and psychopathology. *Development & Psychopathology, 12*, 297–312.

Sarason, I. G., Pierce, G. R., & Sarason, B. R. (1994). General and specific perceptions of social support. In W. R. Avison & I. H. Gotlib (Eds.), *Stress and mental health: Contemporary issues and prospects for the future* (pp. 151–177). New York: Plenum.

Schafer, J. L., & Graham, J. W. (2002). Missing data: Our view of the state of the art. *Psychological Methods, 7*, 147–177.

Simons, R. L., & Associates. (1996). *Understanding differences between divorced and intact families*. Thousand Oaks, CA: Sage.

Singer, M. I., Anglin, T. M., Song, L. Y., & Lunghofer, L. (1995). Adolescents' exposure to violence and associated symptoms of psychological trauma. *JAMA: Journal of the American Medical Association, 273*, 477–482.

Springer, C., & Padgett, D. K. (2000). Gender differences in young adolescents' exposure to violence and rates of PTSD symptomatology. *American Journal of Orthopsychiatry, 70*, 370–379.

Steinberg, L., & Morris, A. S. (2000). Adolescent development. *Annual Review of Psychology, 52*, 83–110.

Stouthamer-Loeber, M., Loeber, R., Homish, D. L., & Wei, E. (2001). Maltreatment of boys and the development of disruptive and delinquent behavior. *Development & Psychopathology, 13*, 941–955.

Sutker, P. B., Davis, J. M., Uddo, M., & Ditta, S. R. (1995). War zone stress, personal resources, and PTSD in Persian Gulf War returnees. *Journal of Abnormal Psychology, 104*, 444–452.

Tiet, Q. Q., Bird, H. R., Hoven, C. W., Moore, R., Wu, P., Wicks, J., et al. (2001). Relationship between specific adverse life events and psychiatric disorders. *Journal of Abnormal Child Psychology, 29*, 153–164.

Trickett, P. K., Reiffman, A., Horowitz, L. A., & Putnam, F. W. (1997). Characteristics of sexual abuse trauma and the prediction of developmental outcomes. In S. L. Toth (Ed.), *Developmental perspectives on trauma: Theory, research, and intervention. Rochester symposium on developmental psychology* (Vol. 8, pp. 289–314). Rochester, NY: University of Rochester Press.

Turner, R. J. (1999). Social support and coping. In A. V. Horwitz & T. L. Scheid (Eds.), *A handbook for the study of mental health: Social contexts, theories, and systems* (pp. 198–210). New York: Cambridge University Press.

Tyano, S., Iancu, I., Solomon, Z., Sever, J., Goldstein, I., Touviana, Y., et al. (1996). Seven-year follow-up of child survivors of a bus-train collision. *Journal of the American Academy of Child & Adolescent Psychiatry, 35*, 365–373.

Ullman, S. E., & Filipas, H. H. (2001). Predictors of PTSD symptom severity and social reactions in sexual assault victims. *Journal of Traumatic Stress, 14*, 369–389.

Wittchen, H. U. (1994). Reliability and validity of the WHO Composite International Diagnostic Interview (CIDI): A critical review. *Journal of Psychiatric Research, 28*, 57–84.

Wolfe, D. A., & Wekerle, C. (1997). Pathways to violence in teen dating relationships. In S. L. Toth (Ed.), *Developmental perspectives on trauma: Theory, research, and intervention. Rochester sym-*

posium on developmental psychology (Vol. 8, pp. 315–341). Rochester, NY: University of Rochester Press.

Wright, M. O. D., Masten, A. S., & Hubbard, J. J. (1997). Long-term effects of massive trauma: Developmental and psychobiological perspectives. In S. L. Toth (Ed.), *Developmental perspectives on trauma: Theory, research, and intervention. Rochester symposium on developmental psychology* (Vol. 8, pp. 181–225). Rochester, NY: University of Rochester Press.

Zanarini, M. C., & Frankenburg, F. R. (1997). Pathways to the development of borderline personality disorder. *Journal of Personality Disorders, 11*, 93–104.

Zoroglu, S. S., Tuzun, U., Sar, V., Tutkun, H., Savas, H. A., Ozturk, M., et al. (2003). Suicide attempt and self-mutilation among Turkish high school students in relation with abuse, neglect and dissociation. *Psychiatry & Clinical Neurosciences, 57*, 119–126.

RESEARCH IN HUMAN DEVELOPMENT, *1*(4), 291–326

Markers of Resilience and Risk: Adult Lives in a Vulnerable Population

J. Heidi Gralinski-Bakker and Stuart T. Hauser
Harvard Medical School
Judge Baker Children's Center

Cori Stott and Rebecca L. Billings
Judge Baker Children's Center

Joseph P. Allen
University of Virginia

In this report, we drew on data from an ongoing longitudinal study that began in 1978 (Hauser, Powers, Noam, Jacobson, Weiss, & Folansbee, 1984). Focusing on late, young-adult life among individuals who were psychiatrically hospitalized during adolescence, we examined markers of resilience empirically defined in terms of adult success and well-being. The study includes a demographically similar group recruited from a public high school. Major goals were to (a) develop preliminary models of adaptive functioning among adults in their 30s, (b) examine the extent to which adults with histories of serious mental disorders can be characterized by these models, and (c) explore predictors of successful adult lives from indicators of individuals' psychosocial adjustment at age 25.

Results showed significant cohort effects on indexes of adaptive functioning, especially for men. Findings suggest that social relations as well as self-views of competence and relatedness play important roles in characterizing adjustment during the adult years. In addition, indexes of psychosocial adjustment as well as symptoms of psychiatric distress and hard drug use at age 25 made a difference in adult social functioning and well-being, providing hints of possible mechanisms likely to facilitate the ability to "bounce back" after a difficult adolescence.

Over the past 20 years, reports in epidemiological, public health, psychological, and psychiatric literatures have reflected a heightened interest in the etiology, pa-

Requests for reprints should be sent to J. Heidi Gralinski-Bakker, Judge Baker Children's Center, Harvard Medical School, 3 Blackfan Circle, Boston, MA 02115. E-mail: jheidi@hms.harvard.edu

thology, and treatment of adolescent mental disorders (U.S. Public Health Service, 1999, 2000). Considerable attention has been devoted to analyses of personal and societal costs of mental disorders including links with adolescents' capacities to carry out their personal, educational, family, and social responsibilities. In general, these analyses have suggested that adolescent-era mental disorders place some individuals at risk for a variety of maladaptive outcomes in their own lives as well as with friends and family, at school, and in the community (Cicchetti & Cohen, 1995). From a developmental perspective, these findings prompted questions about the longer term roles of adolescent mental disorders in adult psychological growth and functioning over time (Gralinski-Bakker, Hauser, Billings, & Allen, 2004; The National Advisory Mental Health Council Workgroup on Child and Adolescent Mental Health Intervention Development and Deployment, 2001).

To address these questions, some recent attention has focused on understanding vulnerabilities that characterize a substantial number of those with a history of mental disorders (U.S. Public Health Service, 2000). This knowledge can be vital in designing intervention programs aimed at identifying risk factors and avoiding undesirable outcomes. At the same time, a growing literature on resilience has suggested that it is equally important to understand positive developmental outcomes among those who have faced significant adversity including the ability of high-risk individuals to "bounce back" in later life after a difficult youth (Garmezy, Masten, & Tellegen, 1984; S. T. Hauser, 1999; S. T. Hauser & Allen, 2004; Masten, Best, & Garmezy, 1990; Rutter, 1987; Werner & Smith, 2001). In particular, knowledge about the ability to bounce back is essential for helping individuals with histories of mental disorders develop satisfying relationships with family and friends, competence in important life roles, and a sense of psychological well-being (Bornstein, Davidson, Keyes, & Moore, 2003). With this interest in mind, our longitudinal study is among the few that have conducted long-term follow-up of adults who were troubled during adolescence (S. T. Hauser, 1999; Vaillant & Vaillant, 1981; Werner & Smith, 2001). However, additional work is needed to build a foundation for understanding capacities to function adaptively in different developmental eras among an especially vulnerable population—those with histories of psychiatric symptoms sufficient to result in psychiatric hospitalization.

In our own research, begun in 1978, we have been interested in development among individuals who experienced psychiatric hospitalization during adolescence. Not only did these individuals suffer from serious psychopathology during an important developmental era, the majority (96%) reported potentially traumatic experiences during their childhood and adolescence. There is little doubt that the general stigma associated with the label of mental illness was made even more deleterious by their living in a psychiatric hospital (S. T. Hauser, 1999; S. T. Hauser & Allen, 2004). In addition, our data have provided evidence of adolescents' relative failure to attain important developmental milestones including performing well in school, acquiring capacities to deal adaptively with stress, developing an increasingly complex and integrated sense of self, and attaining both emotional and behavioral autonomy in relation to parents (Allen, Hauser, Bell, & O'Connor, 1994; S. T. Hauser & Bowlds, 1990).

Despite such evidence of a difficult and troubled youth, Hauser and Allen have drawn on data from this longitudinal study to find instances of positive functioning among some individuals who overcame the adversities they faced in adolescence to live relatively successful lives at age 25 (S. T. Hauser, 1999; S. T. Hauser & Allen, 2004). In this process, Hauser (1999) empirically defined resilience in terms of individual outcome profiles encompassing early adult psychosocial development, relationship functioning, and social competence. Extending this research to participants in a new developmental era (i.e., the "30s"), we now focus on profiles of competence in major life roles that typically define success in late, young-adult lives including those that carry developmental and social significance. In addition to manifest competence in these roles, we consider aspects of psychological well-being especially relevant to adult life. With these interests in mind, there are three overarching goals of the analyses presented in this article: (a) to develop preliminary models of adaptive functioning among adults in their 30s, (b) to examine the extent to which adults with histories of serious mental disorders can be characterized by these models, and (c) to explore predictors of successful adult lives from indicators of individuals' psychosocial adjustment at age 25.

As discussed further below, major life roles for adults in our Western cultural context typically involve work, family, friends, marriage, and parenting (Levinson, Darrow, Klein, Levinson, & McKee, 1978; Levinson & Levinson, 1996). Important aspects of psychological well-being include a sense of overall self-worth as well as an integrated sense of identity and purpose in life (for a recent discussion, see Leary & Tangney, 2003). Psychological well-being also includes beliefs about one's abilities to manage one's life and the surrounding world effectively as well as to develop and sustain meaningful relationships (Keyes & Waterman, 2003; Ryff & Keyes, 1995). In addition to their contributions to well-being, beliefs about competence and relatedness reflect basic psychological needs that are central motives helping to initiate and maintain behavior over time (Baumeister & Leary, 1995; Deci & Ryan, 1991; Ryan & Deci, 2000, 2003). People want to function competently in various life domains; they also want to experience warm, close, and accepting relationships embedded in communal networks. At the same time, those who believe they can manage their lives effectively and maintain supportive relationships have psychological resources for dealing with life challenges (Hobfoll, 2002). Exploring the nature of people's self-views as well as how they function in life domains may therefore provide a meaningful framework to understand critical aspects of adult psychosocial adjustment.

RESILIENCE AS A CONSTRUCT

Recent reviews have emphasized the importance of providing a clear definition of resilience, with particular attention paid to its operationalization in empirical work (Kaplan, 1999; Luthar, Cicchetti, & Becker, 2000). Broadly speaking, resilience has been inferred on the basis of successful adaptation among individuals who faced challenging or threatening circumstances (Luthar et al., 2000; Masten, 2001; Masten et al., 1990; Rutter, 1987). Despite a general consensus about such markers, there is limited agreement about the extent to which external or internal criteria might serve as a basis for defining outcomes as markers of resilience (Luthar, 1999; Luthar et al., 2000; Masten, 2001). For some researchers, manifest competence (such as positive peer relations or freedom from symptoms of substance abuse) is an important indicator of resilient functioning (Rutter, 2000). Others have suggested that internal criteria (such as psychological well-being or low levels of distress) are important determinants of the degree to which individuals are "doing well" in the context of risk (Ryff & Singer, 2003). Recently, Luthar and her colleagues (2000) emphasized the importance of a balanced interpretation of both external and internal criteria. Following some adverse experiences, manifest competence despite some underlying distress may reflect successful adjustment. On the other hand, unremitting distress can jeopardize a person's capacity to function adaptively in everyday life and contribute to deterioration in levels of adaptive functioning over time (Kaplan, 1999). Within these perspectives, most researchers agree that it is important to consider adjustment in domains that are socially and psychologically meaningful to a wide range of people who share a set of cultural traditions. At the same time, it is essential to pay special attention to the population being studied and to set reasonable standards as evidence of success (Luthar et al., 2000; Masten, 1999). As Masten (2001) noted, these are highly complex issues that have only recently received empirical attention.

To examine resilience in this study, we take a life-course perspective focusing on outcomes that might reasonably differentiate people who are comparable to the average general population from those who are experiencing difficulties in their lives (Masten, 2001; Masten & Coatsworth, 1995). In addition, we consider emotional and motivational self-evaluative processes likely to contribute to psychological well-being and to guide behavior and development over time (Baumeister & Leary, 1995; Baumeister & Vohs, 2003; Deci & Ryan, 1991, 2000; Epstein, 1991). In these contexts, we examined the absence of negative outcomes as well as the presence of positive outcomes. For researchers who have focused on the absence of negative outcomes, low levels of impairment typically translate to positive outcomes and therefore to indicators of resilience under conditions of risk (Rutter, 1987, 2000). From this perspective, markers of resilience may include freedom from social deviance or psychopathology such as substance abuse, crime, or psychiatric distress. For participants in this study, we hypothesized that the absence of diagnoses of psychiatric disorders during the past year was such an indicator.

At the same time, an emphasis on the absence of negatives does not necessarily signify the presence of positives (Antonovsky, 1987). People who are free from symptoms of dysfunction may not have developed the competencies needed to address adequately their personal, social, and familial needs. They may not be functioning well at work, at home, or in the community. In addition, they may not be satisfied with the lives they lead or perceive themselves as people of worth. As a result, focusing merely on the absence of negative outcomes tells us little about those who successfully adjust to the challenges of adult roles and responsibilities, achieve long-term favorable outcomes, and live fulfilling adult lives (Bornstein et al., 2003; Seligman & Csikszentmihalyi, 2000).

COMPETENT FUNCTIONING

For the most part, competence in major life roles becomes increasingly significant when adults are in their 30s. Markers of success are typically reflected in manifest indicators of work performance, supportive relationships with family and friends, intimate relationships with significant others, and childbearing (Levinson et al., 1978; Levinson & Levinson, 1996). In addition to being at risk for a range of mental health problems, a substantial minority of those with a history of serious mental disorders is at high risk of failing to attain success in these roles. Evidence of such risk comes from a number of studies that have examined community samples for symptoms of psychiatric disorders. These data suggest that people with a history of mood disorders are at increased risk of repeated unemployment, poor work performance, early or unplanned parenthood, and problematic relationships with romantic partners (Fergusson & Woodward, 2002). Relative to their peers, those with histories of antisocial behavioral disorders have poor work histories in low-status, unskilled jobs and a trend toward early, albeit unstable, marriages and youthful childbearing (Sampson & Laub, 1993). Facing the possibility of such failures, the avoidance of such undesirable outcomes might indeed translate into relative success for some adults with histories of serious mental disorders. At the same time, achievements comparable to the average general population may be even more noteworthy (Masten, 2001), suggesting a metric to be applied to our criteria for evaluating adult success.

Based on our understanding of resilient outcomes among our participants at age 25 (when 13.3% of the hospital group fit empirically defined profiles of resilience; S. T. Hauser, 1999; S. T. Hauser & Allen, 2004), we predicted that a small minority of the previously hospitalized group would show evidence of competent functioning in domains critical to adult success. In contrast, we hypothesized that the presence of severe mental illness in adolescence is sufficiently disruptive of normal developmental processes as to produce lifelong deficits in adult role functioning for a significant minority of those who required psychiatric hospitalization. Because interpersonal relationships are an important element of work life, we predicted further that those adults who experience difficulty in social relations would show evidence of impaired functioning in the work role. Based on previous research that suggested that men tend to be vulnerable to risk conditions during this developmental era (Werner & Smith, 2001), we also hypothesized that men from the hospital group would be at greatest risk for poor outcomes.

PSYCHOLOGICAL WELL-BEING

In modern Western societies, multiple aspects of personal well-being gain importance during the adult years. Consistent with a multidimensional model of well-being (Keyes & Waterman, 2003; Ryff & Keyes, 1995), these include self-evaluations of overall worth, a sense of meaning and purpose in life, perceived efficacy or beliefs about one's abilities to manage life effectively, and beliefs about one's acceptance and involvement in social relationships. There is little doubt that positive evaluations about one's own life and sense of purpose are critical dimensions of subjective well-being (Diener, 1984; Keyes & Waterman, 2003). Not only are perceived efficacy and belonging important aspects of well-being, they are important components of a motivational perspective on psychological needs that guide behavior and development over time and play a significant role in the organization and regulation of people's everyday lives (Deci & Ryan, 1985, 1991; Ryan & Deci, 2003). Broadly speaking, proponents of this perspective have suggested that people are motivated to develop or maintain a perception of themselves as competent, which involves feeling capable of performing goal-directed behaviors (Deci & Ryan, 1985, 1991; White, 1959). People presumably also have a strong need to affiliate or belong as reflected in having a few close, mutually caring, and supportive relationships with others (Baumeister & Leary, 1995; Deci & Ryan, 2000). Moreover, empirical support for links between adult well-being and the important roles of perceived competence (or agency) and relatedness is overwhelming (for a recent review, see Ryan & Deci, 2001).

With respect to a competence motive, people who believe they are capable are more likely to behave effectively in many areas of life (for a recent discussion, see Baumeister & Vohs, 2003). In contrast, those who have doubts about their abilities are likely to view many aspects of their adult lives as full of challenges that are difficult to achieve. For example, beliefs about personal abilities influence job performance and occupational success (Bandura, 1997). When people have confidence in their abilities, they are more likely to persist in the face of work-related challenges than those who doubt their abilities to do well and anticipate the futility of their efforts. In a similar vein, people's beliefs about their abilities to develop and maintain interpersonal relationships make a difference in the relationships they have with others (Caprara et al., 1998). When people believe that they are socially competent, they tend to engage effectively with friends and intimate partners. Similarly, those who have developed and maintained close, mutually supportive relationships are likely to experience themselves as capable of future social success. In addition to satisfying an individual's need to belong, such relationships can be sources of support in challenging circumstances. For those who believe their social skills are limited, however, it may be difficult to form and maintain mutually satisfying relationships. As a result, experiences of acceptance and connectedness to others may be restricted, and potential sources of social support may be lacking (Baumeister & Leary, 1995).

Because links from adolescent-era serious mental disorders and aspects of adult psychological well-being have gone almost completely unexamined to date, our hypotheses are speculative. For some adults with histories of serious mental disorders, positive self-views may indeed reflect overall satisfaction with themselves and healthy well-being (Keyes & Waterman, 2003). Alternatively, overly positive self-views might be considered defensive and potentially maladaptive. For adults who are experiencing difficulties in their lives, such positive self-views may mask less conscious self-doubts and feelings of inadequacy (Brown & Bosson, 2001; Kohut, 1971). They may also reflect an assortment of maladaptive techniques such as pervasive self-serving biases and the denial of responsibility for failure (Kernis & Paradise, 2002). Thus, we hypothesized that positive self-views in concert with evidence of competent functioning in major life roles were likely indicators of well-being. On the other hand, we reasoned that positive self-views in the absence of indicators of adult success were less likely indicators of mental health.

Finally, we considered the role played in successful adult development by indexes of psychosocial adjustment at age 25. Drawing on the research with young adults in our sample (S. T. Hauser, 1999; S. T. Hauser & Allen, 2004), we examined both indicators of adult psychosocial development, relationship functioning, and social competence as well as indicators of social deviance and psychopathology. The first set of indicators included psychological maturity (as reflected in scores of ego development; Loevinger, 1976), ego resiliency as judged by friends (Kobak & Sceery, 1988), self-worth, perceived competence, and perceived sociability (Messer & Harter, 1989)[1] and attachment representation coherence (Main & Goldwyn, 1998). The second set of indicators included hard drug use and criminal behavior in the past 6 months (Elliott, Ageton, Huizinga, Knowles, & Canter, 1983) and total symptoms of psychiatric distress (Derogatis,

[1]In previous work, Hauser (1999) examined indexes of relationship closeness. Because the measures of self-worth, perceived competence, and perceived sociability were conceptually similar to the indexes of psychological well-being included in this study, we included them in the analyses presented in this article.

1983). Given relatively robust findings in research on development in low-risk populations, we predicted that relatively high levels of the psychological resources at age 25 would be associated with markers of successful adult functioning and well-being in the 30s. We also predicted that relatively high levels of social deviance and psychiatric distress would be associated with diminished success.

METHOD

Participants

In this study, we included 118 adults (65 female, 53 male) whose data were complete.[2] These adults represent 83% of the living members ($n = 142$) of an original sample of adolescents first enrolled between 1978 and 1980 in an ongoing longitudinal study of individual and familial development over time (Allen, Hauser, & Borman-Spurrell, 1996; S. T. Hauser, Jacobson, Noam, & Powers, 1983; S. T. Hauser et al., 1984). The original sample included two groups: adolescents who had a nonpsychotic, nonorganic impairment serious enough to require hospitalization ($n = 70$, M age = 14.4 years) and adolescents drawn from 250 volunteers in the freshman class of a public school ($n = 76$, M age = 14.2 years). Participants in this study included 49 members of the original hospital group (26 women, 23 men) and 69 members of the original school group (39 women, 30 men). At the time of the adult assessments, the participants ranged in age from 26 to 35 years, with an average age of 30.35 ($SD = 2.26$) and 31.10 years ($SD = 1.24$) in the hospital and school groups, respectively.

As reported in previous work by Hauser and colleagues (Allen et al., 1996; S. T. Hauser et al., 1983), adolescents from the hospital and school groups did not differ significantly in terms of age, gender, or ethnicity. Although both groups came from families in the middle-class to upper middle-class range, approximate family social status based on Hauser–Warren coefficients (R. M. Hauser & Warren, 1997) indicated significant group differences, $F(1, 144) = 13.50$, $p < .001$, with mean status levels lower in the hospital sample ($M = 44.24$, $SD = 13.90$) than in the school sample ($M = 56.44$, $SD = 17.92$).[3]

Originally, the sampling procedure was used to examine adolescents across a broader range of levels of social functioning than would typically be available in a representative community sample. Psychiatric hospitalization was thus used as a criterion to obtain a sample likely to be at lower levels of individual and family

[2]Data were collected from 9 additional original participants. Because the protocol was not complete, data from these participants were excluded from this study.

[3]Given these significant differences, adolescent-era family social class has been entered into analyses where it is theoretically plausible that adult outcomes (e.g., educational attainment, occupational prestige) may be confounded with or explained by family social class differences.

functioning (for a more complete description of sampling procedures and rationale, see S. T. Hauser, 1991). Initially, the hospitalized adolescents were given diagnoses based on the second edition of the *Diagnostic and Statistical Manual of Mental Disorders* (*DSM–II*; American Psychiatric Association, 1968). In 1992, we updated these diagnostic classifications using criteria from the revised third edition of the *DSM* (*DSM–III–R*; American Psychiatric Association, 1987). Based on a review of the hospital charts (including previous diagnoses), these adolescents' psychiatric symptoms were classified as being conduct disorders (46%), depression and other mood disorders (25%), anxiety disorders (10%), and diverse other classifications (18%).

Preliminary comparison of this follow-up sample with the original sample showed no significant differences in terms of mean current age or percentage male. In the hospital group, those in the follow-up and original samples did not differ in terms of the percentage receiving particular diagnoses as part of the hospital admission process. Within the follow-up sample, the adult lives of participants from the hospital and school groups differed along several developmentally salient dimensions (see Table 1). By their 30s, only 9% of the hospital group had received baccalaureate or advanced degrees, whereas more than half of the school group had completed 4 years of university (43%) or advanced graduate education

TABLE 1
Rates of Participant Educational Attainment, Occupational Status, and Family Structure

	Total[a]		*Men*[b]		*Women*[c]	
Variable	*Hospital*	*School*	*Hospital*	*School*	*Hospital*	*School*
Educational attainment						
Less than high school	.14	.01	.19	.03	.04	.00
High school only	.23	.03	.24	.03	.23	.03
Some postsecondary	.53	.25	.43	.33	.64	.18
Bachelor's degree	.07	.43	.10	.37	.07	.47
Advanced degree	.02	.28	.05	.23	.00	.32
Occupational status						
Imprisoned	.06	.00	.13	.00	.00	.00
Unemployed	.08	.03	.09	.03	.08	.03
Employed	.76	.88	.78	.97	.73	.82
Homemaker	.10	.06	.00	.00	.19	.10
Student	.00	.03	.00	.00	.00	.05
Family structure						
Single	.31	.36	.43	.50	.19	.26
Partnered	.43	.56	.39	.40	.46	.69
Divorced/separated	.20	.03	.13	.03	.27	.03
Currently a parent	.42	.23	.26	.16	.57	.28

[a]In the full sample, $N = 118$: hospital $n = 49$, school $n = 69$. [b]For the men, $n = 53$: hospital $n = 23$, school $n = 30$. [c]For the women, $n = 65$: hospital $n = 26$, school $n = 39$.

(28%). Consistent with extensive prior research, participants' educational attainment was associated with their occupational success, $r = .61$ and $.60$, $p < .001$ for the hospital and school group, respectively. Those employed outside of the home included 76% and 88% of the hospital and school groups, respectively. Of those employed, members of the hospital group, on average, held jobs with lower occupational prestige ($M = 34.04$, $SD = 12.84$) than the jobs held by members of the school group ($M = 48.23$, $SD = 16.61$), $F(1, 95) = 19.49$, $p < .001$. Occupational prestige did not differ by gender or the association between gender and previous psychiatric status. Moreover, the effect of psychiatric status on educational attainment and occupational prestige remained significant after controlling for the effect of adolescent era family social class. Additional comparisons of relationship and childbearing status showed that members of the hospital group were less likely to be in committed relationships, more likely to be separated or divorced, and more likely to have experienced early childbearing than were members of the school group.

Procedures

When participants were in their 30s, on average, they took part in the "adult era" of the larger investigation. Consistent with the informed consent procedures, the adult era included a 6- to 8-hr battery of assessments conducted in private rooms either at our research site or for those participants living at a distance who preferred not to travel, in private rooms at universities near their residences. In general, these assessments included an array of observations, structured and semi-structured interviews, and questionnaires covering a broad range of domains of functioning as well as a psychiatric assessment. As part of this process, participants provided information about their educational attainment and occupational status. They took part in a structured clinical interview focusing on psychiatric symptoms as well as an interview asking about their behavior, feelings, and satisfactions in five role areas: work, social and leisure, extended family, marriage, and parenting (Weissman & Paykel, 1974). In addition, they completed a questionnaire describing their self-views (O'Brien & Epstein, 1987).

In the "early adult phase" of the larger investigation, 142 participants (50% women, M age = 25.8 years) gave their informed consent to participating in a 4-hr battery of assessments focused on developmentally and socially salient domains. Assessments were conducted in private rooms at our research site or at universities within geographic proximity to participants' residences (for a more complete description, see Allen et al., 1996). Of relevance to this study, measures of psychosocial functioning included indexes of ego development, representations of early emotionally significant relationship experiences, self-worth, self-perceptions of competence and sociability, and self-reported hard-drug use, criminality, and symptoms of psychological distress. Using the California Q-sort (Block, 1978), two peers also provided ratings of participants' ego functioning.

Measures

Participant characteristics, adult era. Information about education, employment, relationship status, and childbearing was obtained from a structured interview. Scores for educational attainment were derived from criteria commonly used in developmental research (Entwistle & Astone, 1994). These scores ranged from 0 to 4 reflecting "less than school," "school diploma or equivalent," "some postsecondary education," "university baccalaureate degree," and "graduate or professional degrees." Scores for occupational success were based on ratings assigned by coders to the jobs described by participants. These ratings were given assigned coefficients drawn from the Hauser–Warren total-based socioeconomic index (SEI) (R. M. Hauser & Warren, 1997).

Psychopathology, adult era. Information about current symptoms of diagnostic relevance were assessed with the Structured Clinical Interview for *DSM–III–R* (Spitzer, Williams, Gibbon, & First, 1990). This interview is designed to approximate the differential diagnostic process performed by an experienced clinician. Research assistants who had not been given any information about participants or their psychiatric history conducted the interviews. They were carefully selected, rigorously trained, and closely supervised by expert clinicians. A clinical psychologist or board certified psychiatrist with extensive diagnostic experience reviewed all interviews and diagnoses in research diagnostic conferences. It was the "expert" diagnoses that were used in this study.

Role performance, adult era. A semistructured interview, the Social Adjustment Scale (Weissman, 1995; Weissman & Paykel, 1974), was used to assess five major areas of functioning: work, social and leisure activities, relationships with extended family, marital role, and parenting. The interview asks questions that fall into four major categories: instrumental role performance; the amount of friction with others; interpersonal behaviors such as reticence, withdrawal, and dependency; and feelings such as disinterest or distress. The validity and reliability of the interview has been well demonstrated (Weissman, 1993). Responses to the individual questions were scored using a 5-point scale ranging from 1 to 5 that, in the interviewer's opinion, characterizes the quality of functioning expressed. The first point is used to reflect excellent status, the second point is considered an average rating for the general population, and the remaining three points indicate increasing degrees of impaired functioning. A second set of scores was used to characterize judgments about overall functioning in each role area as well as overall adjustment. These judgments are reflected in scores ranging from 1 to 7, with higher scores indicating a greater degree of impairment.

Research assistants who had not been given any information about participants or their psychiatric history conducted interviews. They were trained to reliability by J. Heidi Gralinski-Bakker who had first established interrater reliability (ranging from 65% to 80%) using data collected as part of a study assessing the discriminant validity of the Adult Attachment Interview (Crowell et al., 1996). Before gathering data used in the current study, interviewers conducted practice interviews and achieved satisfactory interrater agreement, averaging 80% for the individual items and 73% for the global scores. For this study, scores were reversed so that higher scores reflect more competent functioning than do lower scores; only global scores were included in analyses. Participants met the criterion of overall competent functioning in the areas of work, social relations, and extended family relations[4] when they received a global rating indicating excellent or average functioning for the general population.

Self-evaluations, adult era. The 116-item Multidimensional Self-Esteem Inventory (O'Brien & Epstein, 1987) was used to provide measures of psychological well-being. Using a 5-point scale ranging from "almost never" (1) to "very often" (5), participants evaluated their feelings of self-esteem and a sense of identity including a tendency to know what one wants out of life and to have clear long-term goals. Participants also described themselves in terms of their perceived competence and self-control as well as their perceptions of likeability by peers and beliefs about their abilities to love and be loved. Individual scores were created by summing the ratings given to each relevant item. Each score ranged from 10 to 50, with higher scores reflecting more positive self-perceptions than lower scores.

To determine whether these scores were in the high (+2, +1), normal (0), or low (–1, –2) range, participants' ratings on each dimension were transformed into *T* scores using indexes developed by O'Brien and Epstein (1987). For each *T* score at or above 0 (indicating self-ratings at least within the normal range), the participant was considered to have a positive self-view in that domain. Judgments about positive self-views were then organized into three categories. Consistent with a hierarchical model of the self, the first category included evaluations of global self-esteem as well as a sense of identity and purpose. The other two categories included self-perceptions of agency (as reflected in personal ratings of competence and self-control) and perceived relatedness (involving ratings of acceptance and involvement in peer relationships as well as the capacity to express and receive feelings of love).

---

[4]Because a relatively large percentage of participants were not in a marital or parenting role, these indexes were not included.

Ego development, early adult phase. The assessment of ego development was conducted utilizing a 36-item sentence completion test and theoretically derived scoring system (Hy & Loevinger, 1996; Loevinger & Wessler, 1970). Individual differences in ego development represent varying degrees of complexity, openness, and depth in ways that individuals conceptualize self and interpersonal experiences (S. T. Hauser, 1976, 1993; Hy & Loevinger, 1996; Loevinger, 1976). In this study, item sum scores were used to best represent young adults' stage of ego development, with higher scores reflecting increasing psychosocial maturity.

Self-worth and self-perceptions of competence, early adult phase. These self-evaluations were assessed using the Adult Self-Perception Profile (Messer & Harter, 1986), a 50-item scale that taps self-reported ratings of overall self-worth as well as multiple dimensions of an individual's self-perceptions. In addition to overall self-worth, relevant self-ratings used in this study included perceived competence and sociability.

Coherence of representations of early relationships, early adult phase. The Adult Attachment Interview was used to assess representations of early childhood relationships with parents and the possible influence of those representations on personality and development. In addition to overall classifications, the scoring system includes specific scales regarding current states of mind and discourse style (Main & Goldwyn, 1998). In this study, we included the coherence score reflecting individuals' overall ability to organize, integrate, and present a clear, convincing picture and evaluation of past attachment experiences. All interviews were rated by a coleader of the Adult Attachment Institute Workshops (E. Hesse); based on 21 transcripts, interrater reliability for the coherence score was .72. The validity and reliability of the interview and scoring system are well documented (Cassidy & Shaver, 1999).

Peer-rated ego resiliency, early adult phase. Each participant named two peers who were described as "knowing him or her well." These peers were then contacted and asked to rate participants using the California Q-sort (Block, 1978). Peers' ratings were averaged together, and from these ratings, a score for ego resiliency was constructed. This score, ranging from −1.00 to +1.00, was obtained by correlating the sort for each young adult with a criterion sort for the maximally ego-resilient individual provided by Block (1978).

Antisocial behavior, early adult phase. Hard-drug use and criminal behavior were measured with an instrument initially validated and normed in a study (Elliott, Huizinga, & Menard, 1989) of a national probability sample of adolescents. Hard-drug use was measured as the total number of instances of illicit use of five classes of hard drugs (heroin, cocaine, hallucinogens, amphetamines, and tranquilizers) in the previous 6 months. Criminal behavior was measured as the total number of times individuals reported engaging in each of 30 nonoverlapping classes of illegal behavior during the previous 6 months. Measures of both hard-drug use and criminal behavior were highly skewed and were thus transformed using logarithmic transformation.

Symptoms of psychological distress, early adult phase. Using a 4-point rating scale on the Symptom Checklist–90–Revised (Derogatis, 1983), participants indicated the degree to which they experienced distress associated with each of 90 items describing symptoms commonly identified by psychiatric patients. The total number of symptoms endorsed was used as an index of psychological distress.

RESULTS

Overview of the Plan of Analyses

The analyses were organized around questions regarding indicators of resilience among adults who had been hospitalized psychiatrically during adolescence. To provide an overview of adjustment in the two cohorts, we first examined the effects of cohort and gender on presence of psychopathology in the past year (see Table 2). We also examined mean level indicators of competent functioning and

TABLE 2
Rates of Psychopathology Reflected in Psychiatric Diagnoses Within the Past Year

	Total		*Men*		*Women*	
Diagnosis	*Hospital*	*School*	*Hospital*	*School*	*Hospital*	*School*
Any diagnosable disorder[a]	.39	.18	.61	.13	.19	.21
Specific types of disorders[b]						
Mood disorders	.14	.33	.13	.17	.16	.44
Psychotic disorders	.06	.00	.07	.00	.00	.00
Substance disorders	.31	.40	.31	.67	.33	.22
Anxiety disorders	.20	.20	.17	.17	.33	.22
Eating disorders	.03	.06	.00	.00	.16	.11
Antisocial personality	.20	.00	.24	.00	.00	.00
Borderline personality	.06	.00	.07	.00	.00	.00

[a]These rates are based on the full sample, $N = 118$: hospital $n = 49$, school $n = 69$. For the men, $n = 53$: hospital $n = 23$, school $n = 30$. For the women, $n = 65$: hospital $n = 26$, school $n = 39$. [b]These rates include only those who received diagnoses, $n = 50$: hospital $n = 35$, school $n = 15$. For the men, $n = 35$: hospital $n = 29$, school $n = 6$. For the women, $n = 15$: hospital $n = 6$, school $n = 9$.

psychological well-being using multivariate analyses of variance (MANOVAs) to reduce the probability of Type I errors. Significant multivariate findings ($p < .05$) were examined with follow-up univariate analyses of variance (ANOVAs). Table 3 presents the means and standard deviations of these variables as well as results of the univariate significance tests.

Because individual patterns of adjustment are the focus of this report (and they are concealed too easily in group means), we then used loglinear multiway frequency analyses to develop models of competent functioning and well-being for the complete sample. In exploratory analyses, we then applied these to each cohort as a whole and then separately for men and women by cohort. In the final analysis, we examined canonical correlations focusing on patterns of association between early adult indexes of psychosocial adjustment and indexes of adult competent functioning and psychological well-being.

Indicators of Current Psychopathology

Most of the participants did not meet criteria for psychiatric diagnoses within the past year (61% and 82% in the hospital and school groups, respectively). However, these group frequencies obscure significant differences between men who were previously hospitalized and those from the school group: 39% of the men from the hospital group did not receive diagnoses of current psychopathology compared with 87% of the men from the school group. In contrast, the percentages of hospital and school women without diagnoses did not differ (hospital = 81%, school = 79%).

Table 2 shows the overall rates of current psychopathology in the sample as well as rates of different diagnostic categories that characterized their psychiatric symptoms. These data show that mood, substance, and anxiety disorders tended to be relatively common among those with diagnoses, irrespective of group or gender. At the same time, 31% of the hospital men who reported current psychopathology received diagnoses of personality disorders including antisocial behavior as well as borderline personality disorders (24% and 7%, respectively).

Cohort and Gender Effects on Mean Level Indicators of Functioning and Self-Views

Competent Functioning

A 2 × 2 multivariate ANOVA was used to examine cohort and gender effects on global judgments of functioning. These analyses showed only significant multivariate cohort effects, $F(3, 112) = 13.00$, $p < .001$. Follow-up MANOVAs revealed significant cohort differences in functioning in all role areas, with adults from the hospital group functioning at lower levels on average when compared

TABLE 3
Mean Level Indicators of Competent Functioning and Psychological Well-Being

	Total					Men				Women			
	Hospital		School			Hospital		School		Hospital		School	
	M	*SD*	*M*	*SD*	*F(1, 116)*	*M*	*SD*	*M*	*SD*	*M*	*SD*	*M*	*SD*
Competent functioning[a]													
Work	5.33_a	1.81	6.13_b	0.95	09.74*	4.83	2.06	6.17	1.15	5.77	1.48	6.10	0.79
Social relations	5.04_a	1.35	5.99_b	0.81	22.32**	4.74	1.42	5.93	0.83	5.31	1.26	6.03	0.81
Family relations	5.38_a	1.20	6.23_b	0.79	21.16**	5.30	1.15	6.43	0.63	5.46	1.27	6.08	0.87
	M	*SD*	*M*	*SD*	*F(2, 115)*	*M*	*SD*	*M*	*SD*	*M*	*SD*	*M*	*SD*
Psychological well-being[b]													
Self-regard	32.60_a	7.41	35.62_a	6.01	02.01	30.93	7.88	35.80	6.36	34.08	7.25	35.47	5.82
Agency	35.45_a	5.58	37.56_a	4.98	02.35	35.30	5.26	38.68	4.62	35.58	5.95	36.70	5.13
Relatedness	33.97_a	6.18	37.90_b	5.27	08.79*	32.67	5.07	37.01	6.24	35.11	6.91	38.58	4.35
	M	*SD*	*M*	*SD*	*F(3, 114)*	*M*	*SD*	*M*	*SD*	*M*	*SD*	*M*	*SD*
Defensive self-enhancement[c]	46.97	8.38	50.30	6.05	09.10*	43.02_a	6.72	49.43_b	6.43	50.45_b	8.26	50.97_b	5.64

Note. Different subscripts indicate that means are significantly different from each other.

[a]Range = 1–7. High scores indicate competent functioning. [b]Numbers reflect least-square means controlling for defensive self-enhancement. Range = 10–50. High scores indicate positive self-views. [c]There was also a significant interaction effect, $F(3, 114) = 5.50$, $p < .05$.

$*p \leq .01$. $**p \leq .0001$.

with the school group. Although there were no significant multivariate gender main effects or interactions, Table 3 provides descriptive information including means and standard deviations for men and women in each cohort.

Psychological Well-Being

The Cohort × Gender MANCOVA examining the set of variables representing participants' self-views revealed significant cohort, $F(3, 111) = 2.96, p < .05$ and gender, $F(3, 111) = 2.74, p < .05$ main effects. For these analyses, we controlled for the effects of defensive self-enhancement. As shown in Table 3, follow-up 2 × 2 (Cohort × Gender) analyses of covariance revealed only a significant cohort effect on participants' perceived relatedness; other differences were not significant after controlling for the effects of defensive self-enhancement. To better understand the role of defensiveness in these self-reports, a Cohort × Gender ANOVA was performed. Results showed a significant interaction effect, $F(1, 114) = 5.50$, $p < .05$, with men from the hospital group reporting lower levels of self-enhancement when compared with men from the school group and women.

Patterns of Adjustment

Competent Functioning

A three-way frequency analysis was performed to develop a hierarchical linear model of global assessments of functioning in role areas. Dichotomous variables analyzed were whether the participants were rated as functioning at least as well as the average general population at work, in social and leisure activities with friends, and with extended family members. All two-way contingency tables provided expected frequencies in excess of five. Stepwise selection by simple deletion of effects produced a model that included all first-order effects and two-way associations. The model had a likelihood ratio, $\chi^2(1, N = 118) = .04, p = .84$, indicating a good fit between observed frequencies and expected frequencies generated by the model. A summary of the model with results of tests of partial likelihood chi-square and loglinear parameter estimates appears in Table 4. Because of a small sample size, we were unable to develop separate models for each group. Therefore, Table 5 shows results of exploratory two-way contingency tables comparing the observed frequencies in the hospital and school groups as a whole and separately for men and women.

As Table 4 shows, most of the participants were functioning well at work (75%), with friends (64%), and with extended family members (70%). In addition, at least half of the sample showed evidence of competent functioning across multiple role areas. Among those who were rated as being competent in the work role, 54% described competent functioning with friends, whereas 57% received

TABLE 4
Summary of Hierarchical Model of Global Assessments of Competent Functioning

Effect	*Partial Association* χ^2	*Parameter Estimate*	*SE*	*Rates of Competent Functioning* (+,+)	(–,–)
First order					
Work	14.50***	–0.44	.12	.75	
Social relations	0.47	–0.08	.12	.64	
Family relations	5.70*	–0.27	.11	.70	
Second order					
Work × Social Relations	6.66**	0.30	.12	.54	.15
Work × Family Relations	3.51	0.22	.12	.57	.13
Social × Family Relations	5.41*	0.25	.11	.51	.17

Note. Overall evaluation of the model was based on the likelihood ratio, χ^2 (1, N = 118) = 0.04, p = .84.

*$p \leq .05$. **$p \leq .01$. ***$p \leq .0001$.

ratings indicating competence with extended family members. About half (51%) were rated as competent in the two interpersonal domains, with friends as well as extended family members. In contrast, a small minority (13% to 15%) showed evidence of impairment at work and in social relations, whereas 17% described deficits in social functioning with friends as well as extended family.

Extending this model to explore possible group and gender effects, Table 5 shows first-order effects indicating that about half of the adults in the hospital group showed evidence of competent functioning at work (61%), with friends (49%), and with family members (49%). At the same time, a majority of those in the school group were rated as functioning well at work (84%), with friends (75%), and with family (84%). Consistent with our hypotheses, the observed frequencies of success in each of these role areas were significantly lower among men who were previously hospitalized when compared with men in the school group. In contrast, women in both groups were rated as relatively competent in all role areas, with only one significant group difference in extended family relations (50% and 77% for the hospital and school women, respectively).

When examining second-order effects, the data showed that about one third of the adults who were previously hospitalized showed evidence of across-role competencies. Despite comprising a substantial minority, these rates of success are in contrast with nearly two thirds of the adults in the school group who showed competencies across multiple roles. This group difference is most striking when comparing the rates of men functioning competently across multiple roles: Less than one third of the men who were previously hospitalized were rated as being com-

TABLE 5
Summary of Rates of Competent Functioning Separately by Group and Gender

	Total			Men			Women		
Effect	*H*	χ^2	*S*	*H*	χ^2	*S*	*H*	χ^2	*S*
First order									
Work	.61	7.88**	.84	.44	13.37**	.90	.77	0.06	.80
Social relations	.49	8.70**	.75	.44	6.10*	.77	.54	2.93	.74
Family relations	.49	16.63***	.84	.48	13.87**	.93	.50	5.05*	.77

	Total (+,+)		Total	Total (–,–)		Men (+,+)		Men	Men (–,–)		Women (+,+)		Women	Women (–,–)	
Effect	*H*	*S*	χ^2	*H*	*S*	*H*	*S*	χ^2	*H*	*S*	*H*	*S*	χ^2	*H*	*S*
Second order															
Work × Social Relations	.37	.67	12.78**	.27	.07	.26	.70	14.05**	.39	.03	.46	.64	2.05	.15	.10
Work × Family Relations	.39	.70	21.49***	.29	.02	.30	.83	18.86***	.39	.00	.46	.59	5.23	.19	.03
Social × Family Relations	.33	.64	21.03***	.35	.04	.30	.70	15.60**	.39	.00	.35	.59	6.80*	.31	.08

Note. H = hospital group; S = school group.
* $p \leq .05$. ** $p \leq .01$. *** $p \leq .0001$.

petent in multiple roles compared with nearly three fourths of the high school men. Moreover, about one third of the previously hospitalized men (39%) did not describe competent functioning across multiple role areas, whereas few men from the school group (0% to 3%) were rated as failing to meet criteria for success in two role areas. Among women, rates of competent functioning both at work and interpersonally were similar in the two groups. However, the occurrence of success in both social and extended family roles was lower for those in the hospital group (35%) compared with those in the school group (59%). Moreover, about one third of the women from the hospital group showed evidence of impairment in both of these interpersonal roles.

Taken together, the indicators of success in adult roles revealed a number of participants who were functioning well in all areas including 29% of the hospital group and 55% of the school group. From the hospital group, those attaining success in all roles included 22% of the men ($n = 5$) and 35% of the women ($n = 9$). The school group included 63% of the men ($n = 19$) and 49% of the women ($n = 19$).

Psychological Well-Being

A three-way frequency analysis was performed to develop a hierarchical linear model of psychological well-being. Dichotomous variables analyzed were whether the participants reported overall positive self-regard, agency, and relatedness. All two-way contingency tables provided expected frequencies in excess of five. Stepwise selection by simple deletion of effects produced a model that had a likelihood ratio, $\chi^2(1, N = 118) = 1.23$, $p = .27$, indicating a good fit. The model included all first-order effects and two-way associations.

In the summary presented in Table 6, the model showed that most of the study participants reported relatively positive self-views reflected in ratings of positive self-regard, agency, and relatedness. Consistent with a hierarchical model of well-being, most participants also reported positive self-views across various self-aspects. Of those who characterized themselves as having relatively high self-regard, 79% also described themselves as being agentic and 71% described themselves as having a sense of belonging or relatedness. A substantial majority (76%) who described themselves as agentic also described themselves as having close, supportive relationships with others.

When exploring the model applied separately to the two groups, Table 7 shows that a majority of the members of both groups characterized themselves as having positive self-views. Relatively high proportions of participants in each group also reported multiple positive self-views reflected in the two-way associations of self-regard by agency (hospital = 63%, school = 90%), self-regard by relatedness (hospital = 51%, school = 86%), and agency by relatedness (hospital = 63%, school = 86%). Despite these high proportions, significant cohort differences were evident in the percentages of men who described themselves as having positive self-views across a number of dimensions. Of those men in the hospital group who reported

TABLE 6
Summary of Hierarchical Model of Participants' Psychological Well-Being

Effect	*Partial Association* χ^2	*Parameter Estimate*	*SE*	*Proportion Responding Positively*
First order				
Self-regard	5.19	–0.38	.17	.83
Agency	15.60***	–0.75	.19	.88
Relatedness	0.09	–0.06	.19	.79

Effect	*Partial Association* χ^2	*Parameter Estimate*	*SE*	*Rates of Positive Responses* (+,+)	(–,–)
Second order					
Self-regard × Agency	7.83**	0.51	.18	.79	.08
Self-regard × Relatedness	3.32	0.29	.16	.71	.09
Agency × Relatedness	12.48**	0.67	.19	.76	.09

Note. Overall evaluation of the model was based on the likelihood ratio, $\chi^2(1, N = 118) = 1.23, p = .27$.
$^{**}p \leq .01$. $^{***}p \leq .0001$.

positive regard, 44% also described themselves as agentic, and 35% described themselves as having a sense of relatedness; 57% characterized themselves as having both a sense of agency as well as supportive, accepting relationships. This is in contrast to the substantial majority of men in the school group who rated themselves as having multiple positive self-views. Rates of women having multiple positive self-views did not differ significantly in the hospital and school groups.

Overall, a majority of the participants reported positive self-views across the three aspects of well-being examined in this study. These included 51% and 84% of the hospital and school groups, respectively. Those in the hospital group included 35% of the male (n = 8) and 65% of the female (n = 17) participants. In contrast, the school group included 86% of the male (n = 26) and 82% of the female (n = 32) participants.

Competence in Concert With Well-Being

In the final exploratory analysis, we conducted two-way frequency analyses to examine the extent to which participants in each group showed evidence of competent functioning across the role areas of work and relationships with friends and family as well as psychological well-being. As shown in Figure 1, we examined these rates for each cohort as a whole and then separately for men and women.

TABLE 7
Summary of Rates of Psychological Well-Being Separately by Group and Gender

Effect	*Total*			*Men*			*Women*		
	H	χ^2	*S*	*H*	χ^2	*S*	*H*	χ^2	*S*
First order									
Self-regard	.69	11.11**	.93	.52	11.91**	.93	.85	0.96	.92
Agency	.82	3.39	.93	.74	5.88*	.97	.89	0.03	.90
Relatedness	.65	9.16**	.88	.61	4.68*	.87	.69	4.36*	.90

Effect	*(+,+)*			*(–,–)*		*(+,+)*			*(–,–)*		*(+,+)*			*(–,–)*	
	H	*S*	χ^2	*H*	*S*	*H*	*S*	χ^2	*H*	*S*	*H*	*S*	χ^2	*H*	*S*
Second order															
Self-Regard × Agency	.63	.90	12.30**	.12	.04	.44	.93	16.08**	.17	.03	.81	.87	0.49	.08	.05
Self-Regard × Relatedness	.51	.86	16.65**	.16	.04	.35	.87	15.49**	.22	.07	.65	.85	3.76	.12	.03
Agency × Relatedness	.63	.86	8.36*	.16	.04	.57	.87	6.69*	.22	.03	.69	.85	2.21	.12	.05

Note. H = hospital group; S = school group.
*$p \leq .05$. **$p \leq .01$.

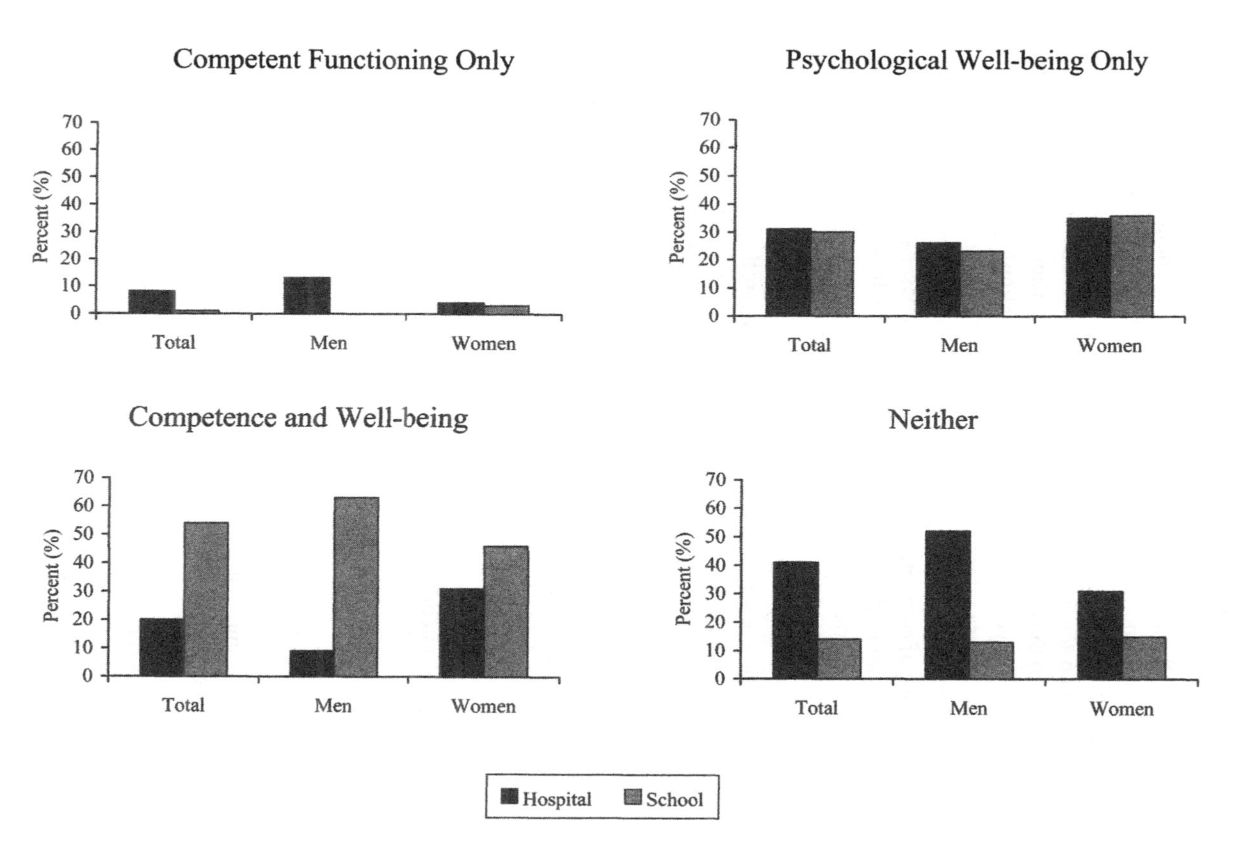

FIGURE 1 Percentage of the full sample with evidence of diverse indicators of resilience.

Separate analyses focused on (a) those who met the criteria of competent functioning only, (b) those who met the criteria of psychological well-being only, (c) those who met both sets of these criteria, and (d) those who did not meet either of these criteria. Overall, these results were consistent with findings reported separately for competent functioning and well-being: Rates of psychosocial adjustment were lower in the hospital group when compared with the school group, men in the hospital group tended to be least successful, and women in the hospital group were similar to those from the school group. In particular, a small minority of the participants showed evidence of competent functioning in the absence of positive self-views. These rates were relatively similar among men and women in each group. An additional 30% to 31% of the participants reported positive self-views in the absence of competent functioning in role areas. Among these, rates of men and women from each group ranged from 23% to 36% with no significant group differences. In contrast, the presence of both competent functioning and well-being showed a significant group effect, $\chi^2(1, N = 118) = 13.19, p < .001$, evident in higher rates of adjustment among members of the school group compared with the hospital group. For the most part, this effect was reflected in significantly different rates of positive outcomes, $\chi^2(1) = 16.25, p < .001$ among men in the hospital group (9%) compared with those in the school group (63%). Similarly, rates of those who appeared to fare poorly revealed a significant group effect, $\chi^2(1) = 10.47, p < .001$. This effect was evident in the relatively high rate of men in the hospital group (52%) who did not show evidence of positive outcomes measured in this study. Rates of positive adjustment as well as the absence of indexes of adjustment did not differ significantly between women in the hospital and school groups.

When absence of current psychopathology was considered in concert with presence of positive outcomes, results showed a significant group effect, $\chi^2(2) = 16.18$, $p < .001$. Those who met all three criteria included 12% of the hospital group (1 man, 5 women) and 46% of the school group (16 men and women).

Early Adult Predictors of Adult Adjustment

A series of four canonical correlation analyses were performed to examine associations between sets of variables reflecting psychosocial adjustment and problematic functioning when participants were 25 years old, on average, and the sets of variables representing adult competent functioning in role areas and psychological well-being. In the first two analyses, we examined patterns of linear associations between adult competent functioning and indexes of early adult psychosocial adjustment and problematic functioning (social deviance and psychiatric symptoms). The second set of analyses examined associations between adult psychological well-being and the sets of variables reflecting early adult psychosocial adjustment and problematic functioning. Results are presented in Tables 8 and 9.

TABLE 8
Summary of Canonical Correlation Analyses Examining Sets of Early Adult Indexes of Psychosocial Adjustment and Adult Indexes of Competent Functioning

	Canonical Variate[a]			Canonical Variate[b]	
Index	*Coefficient*	*Correlation*	*Index*	*Coefficient*	*Correlation*
Early adult set			Early adult set		
Ego development	.22	.55	Psychiatric symptoms	.58	.76
Ego resilience	.30	.69	Hard-drug use	.51	.73
Coherence	.46	.71	Criminality	.29	.62
Global self-worth	.01	.59			
Perceived competence	.20	.52			
Perceived sociability	.37	.64			
Percent of overlapping variance with adult set		.18	Percent of overlapping variance with adult set		.13
Adult set			Adult set		
Work	.21	.54	Work	−.17	−.57
Social relations	.77	.94	Social relations	−.79	−.95
Family relations	.28	.57	Family relations	−.26	−.58
Percent of overlapping variance with early adult set		.25	Percent of overlapping variance with early adult set		.11

[a]Overall evaluation of the presence of canonical variates was based on the likelihood ratio, $F(18, 100) = 2.23, p < .0001$ for one variate. [b]Evaluation of the presence of one canonical variate was based on the likelihood ratio, $F(9, 109) = 3.52, p = .0004$.

TABLE 9
Summary of Canonical Correlation Analyses Examining Sets of Early Adult Indexes of Psychosocial Adjustment and Adult Indexes of Competent Functioning

Index	*Canonical Variate*[a]	
	Coefficient	*Correlation*
Early adult set		
Ego development	.20	.41
Ego resilience	–.11	.35
Coherence	.31	.51
Global self-worth	.31	.83
Perceived competence	.35	.71
Perceived sociability	.38	.78
Percent of overlapping variance with adult set		.17
Adult set		
Self-regard	–.13	.75
Agency	.74	.94
Relatedness	.48	.84
Percent of overlapping variance with early adult set		.31

Index	*Canonical Variate*[b]	
	Coefficient	*Correlation*
Early adult set		
Psychiatric symptoms	.86	.95
Hard-drug use	.12	.38
Criminality	.25	.52
Percent of overlapping variance with adult set		.13
Adult set		
Self-regard	–.62	–.97
Agency	–.15	–.79
Relatedness	–.33	–.87
Percent of overlapping variance with early adult set		.14

[a]Overall evaluation of the presence of canonical variates was based on the likelihood ratio, $F(18, 100) = 4.51, p < .0001$ for one variate. [b]Evaluation of the presence of one canonical variate was based on the likelihood ratio, $F(9, 109) = 2.90, p < .01$.

The first analysis revealed a significant canonical correlation of .50, representing 33% of the overlapping variance for the first pair of canonical variates. As shown in Table 8, for the early adult set, this pair of variates had high loadings on ego resiliency as rated by peers, the coherence score reflecting participants' overall ability to present a believable picture of past attachment experiences and their effects on adult personality and self-reported sociability (.30, .46, and .37, respectively). For the adult set, there was a high loading on social functioning (.77). Thus, high ego resiliency, a coherent discourse style, and perceived sociability at age 25 were related to the quality of adult social functioning.

The second analysis revealed a significant canonical correlation of .46, representing 27% of the overlapping variance for the first pair of canonical variates. Table 8 shows that this pair of variates has high loadings on the presence of psychiatric symptoms and hard-drug use (.58 and .51, respectively) for the early adult set and on social functioning (–.79) for the adult set. Thus, high levels of self-reported psychiatric symptoms and hard-drug use at age 25 were associated with relatively poor adult social functioning.

Results of the third analysis (shown in Table 9) revealed one significant canonical correlation of .65, representing 77% of the overlapping variance for the first pair of canonical variates. For the early adult set, this pair of variates had high loadings on the coherence score characterizing representations of past attachment relationships as well as perceived self-worth, competence, and sociability (.31, .31, .35, and .38, respectively). For the adult set, there were high loadings on perceived agency and relatedness (.74 and .48, respectively). Thus, high levels of attachment coherence and positive self-views at age 25 were related to high levels of adult self-views of agency and belonging.

With respect to association between early adult indicators of difficulty and adult well-being, the analysis revealed a significant canonical correlation of .41, representing 21% of the overlapping variance for the first pair of canonical variates. Table 9 shows that this pair of variates had high loadings on the presence of psychiatric symptoms (.86) for the early adult set and on self-regard and sense of relatedness (–.62 and –.33, respectively) for the adult set. Thus, high levels of self-reported psychiatric symptoms at age 25 were associated with diminished self-regard and feelings of relatedness with others.

DISCUSSION

This study was designed to understand positive outcomes among a group of adults with histories of serious mental disorders during adolescence. In general, we focused on descriptive analyses of development over time among an especially vulnerable population. In particular, we set out to accomplish the following: (a) identify markers of developmentally salient adaptive functioning for adults in their 30s, (b) examine the extent to which adults from an especially vulnerable adolescent population can be characterized by models of adaptive functioning across multiple domains, and (c) explore predictors of successful adult lives from indicators of individuals' psychosocial adjustment at age 25. Despite the relatively common occurrence of serious mental disorders during adolescence (U.S. Department of Health and Human Services, 1999; U.S. Public Health Service, 2000), previous follow-up research of this kind has been limited.

In general, the results of this study highlight the importance of understanding multiple facets of the ability to bounce back among people who had serious adolescent era mental disorders. Our data showed that 61% of these adults did not have diagnoses of current psychopathology, 29% showed evidence of competent functioning across role areas critical to successful adult lives, and 51% characterized themselves as having a sense of psychological well-being. When these indexes were viewed from a person-oriented perspective, however, only 12% of the adults (1 man, 5 women) met all of these criteria in comparison with 46% of the adults (16 men, 16 women) in the group recruited from a suburban high school. The latter person-oriented finding—showing positive outcomes among 12% of the previously hospitalized group—is strikingly similar to our results obtained when participants were 25 years old. Using empirically defined profiles based on criteria salient during the early adult era, Hauser and Allen have found that 13.3% of the previously hospitalized participants were doing relatively well (Hauser, 1999; Hauser & Allen, 2004). These findings draw attention to the difficulties facing adults with histories of serious mental disorders as they struggle to adjust to the demands of adult life. The data collected from adults in their 30s further suggest significant difficulties for varying and sometimes quite high proportions of men who had adolescent era, serious mental disorders.

In general, the findings are consistent with previous large-scale studies about capacities to bounce back among adults who experienced difficulties during childhood and adolescence. Specifically, rates of current psychopathology are similar to findings about long-term (dis)continuities of symptomatology among people with histories of conduct disorder (Robins, 1966) as well as recovery among high-risk youth on the island of Kauai (Werner & Smith, 2001). In addition, the findings suggest that functioning in role areas is most difficult for adults with histories of serious mental disorders. Both men and women in our high-risk sample experienced difficulty in interpersonal relations with friends and extended family. This finding is consistent with data gathered from these participants at age 25 and examined in other reports (Allen & Hauser, 1996; Allen et al., 1996; Allen, Hauser, O'Connor, & Bell, 2002; Allen & Land, 1999; Allen, Moore, Kuperminc, & Bell, 1998). Taken together, the results suggest that the presence of serious mental illness in adolescence may be sufficiently disruptive of normal developmental processes as to produce lifelong deficits in interpersonal functioning. Alternatively, certain forms of earlier psychiatric symptoms might lead to different types of deficits—with mood disorders predicting social withdrawal and isolation and antisocial behavior problems predicting active disengagement from socially sanctioned roles as well as hostility in interpersonal relationships. Clearly, disentangling these alternatives will play an important role in understanding possible effects of serious mental disorders on social development over time.

In addition, previous research suggested that successful performance at work requires the abilities to manage one's interpersonal functioning (Vaillant & Vaillant, 1981). The capacity to have relationships that are characterized by low levels of tension or friction is clearly an important element of interpersonal success in the work and social environs (Elder, 1998). For men from the hospital group, in particular, less than one third reported competent functioning both at work and with interpersonal relations with friends and family. Perhaps, as suggested previously, those who are able to get along well with others in the work environment are more likely to be successful than are those who are disengaged or dysregulated. Some speculative support for this assertion comes from the pattern of diagnostic findings indicating that 61% of the men from the hospital group received diagnoses of current psychopathology. Of these, 31% reported competent functioning at work, 36% had ratings indicating social competence with friends, and 43% had reasonable extended family relations. This leaves nearly two thirds of the men with current psychiatric diagnoses facing difficulty in adult role areas. Such a finding underscores the importance of explicating the roles of past and current psychopathology in men's role deficits. Given the importance of success in adult roles, there also is a need for understanding the mechanisms involved in differentiating men who are able to bounce back from those who experience difficulty.

For women, the rates of those who attained success in major life roles were relatively high—and often comparable to the school sample. At least 50% were doing well in one role area, and 35% to 46% met the criteria indicating success in two roles. As noted previously, women from the hospital group were least successful in social and extended family relations. In contrast with men, however, these social deficits did not appear to take a serious toll on women's capacities to be successful at work. Approximately 75% of the women received ratings of competent functioning at work, and nearly half (46%) were rated as competent at work as well as with friends or extended family. Even though a third of the men and women in the hospital group experienced difficulties in relations with both friends and family, less than 20% of the women had overlapping deficits between interpersonal roles and work. In contrast, nearly 40% of the men experienced overlapping work and interpersonal difficulties. These findings raise questions about the effects of gender on the nature and role of interpersonal relationships in successful adult lives. They also suggest that different mechanisms may underlie role success and role deficits for men and women.

In a similar vein, there were gender differences in the proportions of previously hospitalized men and women who endorsed positive self-views suggesting psychological health or well-being. Those who characterized themselves as having a sense of psychological well-being included 35% of the men and 65% of the women from the hospital group in contrast with 86% and 82% of the school men and women, respectively. In addition to being essential for happiness and a sense of well-being, these positive self-views are likely to serve as psychological resources in dealing with life challenges (Hobfoll, 2002). For example, people who have a sense of agency tend to believe they can manage potentially difficult situations and control their own thoughts, feelings, and behaviors. As a result, they tend to engage actively in new challenges they face and to cope effectively with new demands; they also tend to set reasonable goals and to use their resources effectively (Bandura, 1997; Maddux & Gosselin, 2003). Those with a sense of relatedness or belonging are likely to experience intimacy and support in their relationships (Allen & Land, 1999; Baumeister & Leary, 1995). In this context, they may engage in productive, problem-solving discussions and practice ways of dealing with their emotions; they are also likely to have confidence that their relationships will be enduring yet move freely to establish new relationships.

Overall, about 30% of the adults met the criteria of psychological health even though they experienced difficulties in their life roles. For these adults, positive self-views may indeed reflect healthy well-being because well-adjusted people characteristically maintain positive illusions about themselves (Taylor & Brown, 1988). In this vein, positive self-views have been linked to a number of behaviors that can be viewed as adaptive including realistic optimism, persistence in the face of challenge, and satisfaction with life (Diener, 1984). On the other hand, overly positive or overly negative self-views can be viewed as defensive and potentially maladaptive. In this context, they may reflect self-serving biases aimed either at the denial of responsibility for failure or at protecting the self from disappointments (Kernis & Paradise, 2002). Whether positive self-views among the people in this group reflected healthy well-being or maladjustment merits further attention in our work.

Given our interest in development over time, we had hypothesized links between adult markers of resilience and indexes of early adult psychosocial functioning. In general, the findings support this view; they also highlight the important role of interpersonal relationships, positive self-views, and symptoms of distress. In particular, the abilities to provide a believable account of early attachment relationships and perceived sociability at age 25 predicted the quality of adult social functioning. These findings are consistent with a large body of research that has documented links from representations of early emotionally significant relationships to later social functioning (Cassidy & Shaver, 1999). Coherence and sociability, along with perceived self-esteem and competence, also predicted later perceived agency and relatedness. In addition to the well-established significance of individuals' coherence in describing early caregiver attachment relationships, these patterns of association suggest links with concurrent satisfaction of psychological needs for relatedness and positive self-regard (Allen et al., 1996; Kobak & Sceery, 1988). From a motivational perspective, satisfaction of these needs can be further understood as sources of support enabling subsequent competent functioning and well-being in adult life (Ryan & Deci, 2001).

From these findings, we can only speculate on some of the processes that might be involved. With psychological resources (perceptions of worth, competence, and sociability), some early adults may have been better equipped to cope with the multiple challenges of adult life (Hobfoll, 2002). In particular, people who have high self-worth as well as confidence in their abilities to make friends and to be successful may feel secure in themselves and therefore, more willing to engage socially with others (Bandura, 1997). In contrast, diminished self-views of overall worth, competence, and sociability may have exacerbated the interpersonal challenges facing young adults and contributed to further declines in beliefs about their abilities to manage their lives and have close, supportive relationships.

Along similar lines, relatively high symptoms of psychiatric distress and hard-drug use at age 25 predicted poor social functioning as well as diminished sense of worth and meaning in life and a sense of relatedness with others. These findings provide some suggestion of a pattern of continuity in difficulties over time. For some early adults, psychiatric distress and hard-drug use may have further derailed their development and heightened the risk of social deficits over time. In particular, those who suffer continuing emotional and developmental disruptions may face decreased odds of "catching up" with age appropriate psychosocial development (Allen & Hauser, 1996; Allen et al., 1996). As a result, they are likely to face a heightened risk that their successful adult functioning will be compromised. Further exploration of these links is warranted.

Overall, the findings from this study should be interpreted with caution. The findings are based on small sample sizes and clearly require replication. The small sample size may have constrained the power of some analyses to detect group differences. Among women, in particular, some variations in outcomes that failed to reach conventional statistical significance in our data might emerge as significant in larger groups. The sample population also was limited to adults who came from White, predominantly middle-class and upper middle-class families. This restricted ethnic group and social class membership may limit the generalizability of the findings to other groups. In addition, the need for hospitalization has been used as an indicator of the severity of psychiatric symptoms facing these adolescents as well as a reflection of potentially traumatic experiences before and during hospitalization. A limitation of this study is that specific indexes of trauma have not been included. In addition, adolescent era psychiatric problems among the high school group have not been addressed.

If replicated, however, these findings are important in highlighting the diversity in development among an especially vulnerable population. Our formulation of markers of resilience may provide a meaningful framework for understanding risks facing people who experienced adolescent era, serious mental disorders. The results also suggest a number of directions for future research. First, it will be important to consider a more extensive assessment of positive adjustment using multiple informants and methods rather than a single informant and essentially similar method. Second, it will be important to examine the role of adolescent era psychosocial adjustment in later adult development including patterns of continuity and change in both cohorts. Finally, it is essential that we consider the role of the social context, including aspects of the environment likely to facilitate or impede positive outcomes over time. Within these future directions, a critical element includes identifying mechanisms that might explain the results that were found. We are now pursuing these avenues in our research using additional data gathered in the context of our ongoing longitudinal investigation. At this point, we recognize that the data presented in this study are drawn from interviews and self-reports that reflect the psychological constructions of the participants. Although we cannot be certain of the veracity of these reports, we believe that these findings contribute to an important foundation that can be further developed.

ACKNOWLEDGMENTS

This research was supported by a grant from the National Institute of Mental Health R01 MH44934-12, "Adolescent Paths to Successful Midlife Adjustment." We are grateful to the participants and to the numerous research assistants and associates who made this work possible, including Dr. Robert J. Waldinger, who ensured the accuracy of the diagnostic information. J. Heidi Gralinski-Bakker is also grateful to J. I. ("Hans") Bakker, who very generously gave feedback and support.

REFERENCES

Allen, J. P., & Hauser, S. T. (1996). Autonomy and relatedness in adolescent-family interactions as predictors of young adults' states of mind regarding attachment. *Development and Psychopathology, 3*, 793–809.

Allen, J. P., Hauser, S. T., Bell, K. L., & O'Connor, T. G. (1994). Longitudinal assessment of autonomy and relatedness in adolescent family interactions as predictors of adolescent ego development and self-esteem. *Child Development, 65*, 179–194.

Allen, J. P., Hauser, S. T., & Borman-Spurrell, E. (1996). Attachment theory as a framework for understanding sequelae of severe adolescent psychopathology: An 11-year follow-up study. *Journal of Consulting and Clinical Psychology, 64*, 254–263.

Allen, J. P., Hauser, S. T., O'Connor, T. G., & Bell, K. L. (2002). Prediction of peer-rated adult hostility from autonomy struggles in adolescent-family interactions. *Development and Psychopathology, 14*, 123–127.

Allen, J. P., & Land, D. (1999). Attachment in adolescence. In J. Cassidy & P. R. Shaver (Eds.), *Handbook of attachment: Theory, research, and clinical applications* (pp. 319–335). New York: Guilford.

Allen, J. P., Moore, C. M., Kuperminc, G. P., & Bell, K. L. (1998). Attachment and adolescent psychosocial functioning. *Child Development, 69*, 1406–1419.

American Psychiatric Association. (1968). *Diagnostic and statistical manual of mental disorders* (2nd ed.). Washington, DC: Author.

American Psychiatric Association. (1987). *Diagnostic and statistical manual of mental disorders* (3rd ed., rev.). Washington, DC: Author.

Antonovsky, A. (1987). *Unravelling the mystery of health: How people manage stress and stay well.* San Francisco: Jossey-Bass.

Bandura, A. (1997). *Self-efficacy: The exercise of control.* New York: Freeman.

Baumeister, R. F., & Leary, M. R. (1995). The need to belong: Desire for interpersonal attachments as a fundamental human motivation. *Psychological Bulletin, 111*, 497–529.

Baumeister, R. F., & Vohs, K. D. (2003). Self-regulation and the executive function of the self. In M. R. Leary & J. P. Tangney (Eds.), *Handbook of self and identity* (pp. 197–217): New York: Guilford.

Block, J. (1978). *The Q-sort method of personality assessment and psychiatric research.* Palo Alto, CA: Consulting Psychologists Press.

Bornstein, M. H., Davidson, L., Keyes, C. L. M., & Moore, K. A. (Eds.). (2003). *Well-being: Positive development across the life course.* Mahwah, NJ: Lawrence Erlbaum Associates, Inc.

Brown, R. P., & Bosson, J. K. (2001). Narcissus meets Sisyphus: Self-love, self-loathing, and the never-ending pursuit of self-worth. *Psychological Inquiry, 12*, 210–213.

Caprara, G. V., Scabini, E., Barbaranelli, C., Pastorelli, C., Regalia, C., & Bandura, A. (1998). Impact of adolescents' perceived self-regulatory efficacy on familial communication and antisocial conduct. *European Psychologist, 3*, 125–132.

Cassidy, J., & Shaver, P. R. (Eds.). (1999). *Handbook of attachment: Theory, research, and clinical applications.* New York: Guilford.

Cicchetti, D., & Cohen, D. J. (Eds.). (1995). *Developmental psychopathology.* New York: Wiley.

Crowell, J. A., Waters, E., Treboux, D., O'Connor, E., Colon-Downs, C., & Feider, O. (1996). Discriminant validity of the Adult Attachment Interview. *Child Development, 67*, 2584–2599.

Deci, E. L., & Ryan, R. M. (1985). *Intrinsic motivation and self-determination in human behavior.* New York: Plenum.

Deci, E. L., & Ryan, R. M. (1991). A motivational approach to self: Integration in personality. In R. Dienstbier (Ed.), *Nebraska Symposium on Motivation: Vol. 38. Perspectives on motivation* (pp. 237–288). Lincoln: University of Nebraska Press.

Deci, E. L., & Ryan, R. M. (2000). The "what" and "why" of goal pursuits: Human needs and the self-determination of behavior. *Psychological Inquiry, 11*, 227–268.

Derogatis, L. R. (1983). *SCL–90 adminstration and scoring manual.* Towson, MD: Clinical Psychometric Research.

Diener, E. (1984). Subjective well-being. *Psychological Bulletin, 95*, 542–575.

Elder, G. H. (1998). The life course and human development. In W. Damon (Ed.), *Handbook of child psychology* (Vol. 1, pp. 939–991). New York: Wiley.

Elliott, D. S., Ageton, S. S., Huizinga, D., Knowles, B. A., & Canter, R. J. (1983). *The prevalence and incidence of delinquent behavior: 1976–1980.* Boulder, CO: Behavioral Research Institute.

Elliott, D. S., Huizinga, D., & Menard, S. (1989). *Multiple problem youth: Delinquency, substance use, and mental health problems.* New York: Springer-Verlag.

Entwistle, D. R., & Astone, N. M. (1994). Some practical guidelines for measuring youth's race/ethnicity and socioeconomic status. *Child Development, 65*, 1521–1540.

Epstein, S. (1991). Cognitive-experiential self-theory: Implications for developmental psychology. In M. R. Gunnar & L. A. Sroufe (Eds.), *Self-processes and development* (Vol. 23, pp. 111–137). Hillsdale, NJ: Lawrence Erlbaum Associates, Inc.

Fergusson, D. M., & Woodward, L. J. (2002). Mental health, educational, and social role outcomes of adolescents with depression. *Archives of General Psychiatry, 59*, 225–231.

Garmezy, N., Masten, A. S., & Tellegen, A. (1984). The study of stress and competence in children: A buidling block for developmental psychopathology. *Child Development, 55*, 97–111.

Gralinski-Bakker, J. H., Hauser, S. T., Billings, R. L., & Allen, J. P. (2004). Risks along the road to adulthood: Challenges faced by youth with serious mental disorders. In D. W. Osgood, C. Flanagan, & E. M. Foster (Eds.), *Transitions to becoming an adult among vulnerable populations*. Chicago: University of Chicago Press.

Hauser, R. M., & Warren, J. R. (1997). Socioeconomic index of occupational status: A review, update, and critique. *Sociological Methodology, 27*, 177–298.

Hauser, S. T. (1976). Loevinger's model and measure of ego development: A critical review. *Psychological Bulletin, 33*, 933–955.

Hauser, S. T. (with Powers, S. I., & Noam, G. G.). (1991). *Adolescents and their families: Paths of ego development*. New York: Free Press.

Hauser, S. T. (1993). Loevinger's model and measure of ego development: A critical review. II. *Psychological Inquiry, 4*, 23–30.

Hauser, S. T. (1999). Understanding resilient outcomes: Adolescent lives across time and generations. *Journal of Research on Adolescence, 9*, 1–24.

Hauser, S. T., & Allen, J. P. (2004). *Climbing back: Negotiating a perilous adolescence*. Manuscript under contract.

Hauser, S. T., & Bowlds, M. K. (1990). Stress, coping, and adaptation. In S. S. Feldman & G. R. Elliott (Eds.), *At the threshold: The developing adolescent* (pp. 388–413). Cambridge, MA: Harvard University Press.

Hauser, S. T., Jacobson, A. M., Noam, G. G., & Powers, S. I. (1983). Ego development and self-image complexity in early adolescence. *Archives of General Psychiatry, 40*, 325–331.

Hauser, S. T., Powers, S. I., Noam, G. G., Jacobson, A. M., Weiss, B., & Folansbee, D. J. (1984). Familial contexts of adolescent ego development. *Child Development, 55*, 195–213.

Hobfoll, S. E. (2002). Social and psychological resources and adaptation. *Review of General Psychology, 6*, 307–324.

Hy, L., & Loevinger, J. (1996). *Measuring ego development* (2nd ed.). Mahwah, NJ: Lawrence Erlbaum Associates, Inc.

Kaplan, H. B. (1999). Toward an understanding of resilience: A critical review of definitions and models. In M. D. Glantz & J. L. Johnson (Eds.), *Resilience and development: Positive life adaptations* (pp. 17–83). New York: Kluwer Academic/Plenum.

Kernis, M. H., & Paradise, A. W. (2002). Distinguishing between secure and fragile forms of high self-esteem. In E. L. Deci & R. M. Ryan (Eds.), *Handbook of self-determination research* (pp. 339–360). Rochester, NY: University of Rochester Press.

Keyes, C. L. M., & Waterman, M. B. (2003). Dimensions of well-being and mental health in adulthood. In M. H. Bornstein, L. Davidson, C. L. M. Keyes, & K. A. Moore (Eds.), *Well-being: Positive development across the life course* (pp. 477–497). Mahwah, NJ: Lawrence Erlbaum Associates, Inc.

Kobak, R. R., & Sceery, A. (1988). Attachment in late adolescence: Working models, affect regulation, and representations of self and others. *Child Development, 59*, 135–146.

Kohut, H. (1971). *The analysis of self*. New York: International Universities Press.

Leary, M. R., & Tangney, J. P. (Eds.). (2003). *Handbook of self and identity*. New York: Guilford.

Levinson, D. J., Darrow, C. N., Klein, E. B., Levinson, M. H., & McKee, B. (1978). *The seasons of a man's life*. New York: Knopf.

Levinson, D. J., & Levinson, J. D. (1996). *The seasons of a woman's life*. New York: Knopf.

Loevinger, J. (1976). *Ego development: Conceptions and theories*. San Francisco: Jossey-Bass.

Loevinger, J., & Wessler, R. (1970). *Measuring ego development: Construction and use of a sentence completion test*. San Francisco: Jossey-Bass.

Luthar, S. S. (1999). Measurement issues in the empirical study of resilience: An overview. In M. D. Glantz & J. L. Johnson (Eds.), *Resilience and development: Positive life adaptations* (pp. 129–160). New York: Plenum.

Luthar, S. S., Cicchetti, D., & Becker, B. (2000). The construct of resilience: A critical evaluation and guidelines for future research. *Child Development, 71*, 543–562.

Maddux, J. E., & Gosselin, J. T. (2003). Self-efficacy. In M. R. Leary & J. P. Tangney (Eds.), *Handbook of self and identity* (pp. 218–238). New York: Guilford.

Main, M., & Goldwyn, R. (1998). *Attachment scoring and classification systems*. Unpublished manuscript, University of California, Berkeley.

Masten, A. S. (1999). Resilience comes of age: Reflections on the past and outlook for the next generation of research. In M. D. Glantz & J. L. Johnson (Eds.), *Resilience and development: Positive life adaptations* (pp. 282–296). New York: Kluwer Academic/Plenum.

Masten, A. S. (2001). Ordinary magic: Resilience processes in development. *American Psychologist, 56*, 227–238.

Masten, A. S., Best, K. M., & Garmezy, N. (1990). Resilience and development: Contributions from the study of children who overcome poverty. *Development and Psychopathology, 2*, 425–444.

Masten, A. S., & Coatsworth, J. D. (1995). Competence, resilience, and psychopathology. In D. Cicchetti & D. Cohen (Eds.), *Developmental psychopathology: Vol 2. Risk disorder and adaptation* (pp. 715–752). New York: Wiley.

Messer, B., & Harter, S. (1986). *Adult self-perception manual*. Unpublished manuscript, University of Denver, CO.

Messer, B., & Harter, S. (1989). *The self-perception profile for adults*. Unpublished manual, University of Denver, Denver, CO.

The National Advisory Mental Health Council Workgroup on Child and Adolescent Mental Health Intervention Development and Deployment. (2001). *Blueprint for change: Research on child and adolescent mental health*. Washington, DC: National Institute of Mental Health.

O'Brien, E. J., & Epstein, S. S. (1987). *MSEI: The Multidimensional Self-Esteem Inventory. Professional manual*. Odessa, FL: Psychological Assessment Resources.

Robins, L. N. (1966). *Deviant children grown up*. Baltimore: Williams & Wilkins.

Rutter, M. (1987). Psychosocial resilience and protective mechanisms. *American Journal of Orthopsychiatry, 57*, 316–331.

Rutter, M. (2000). Resilience reconsidered: Conceptual considerations, empirical findings, and policy implications. In J. P. Shonkoff & S. J. Meisels (Eds.), *Handbook of early childhood intervention* (pp. 651–682). Cambridge, England: Cambridge University Press.

Ryan, R. M., & Deci, E. I. (2000). Self-determination theory and the facilitation of intrinsic motivation, social development, and well-being. *American Psychologist, 55*, 68–78.

Ryan, R. M., & Deci, E. L. (2001). On happiness and human potentials: A review of research on hedonic and eudaimonic well-being. *Annual Review of Psychology, 52*, 141–166.

Ryan, R. M., & Deci, E. L. (2003). On assimilating identities to the self: A self-determination theory perspective on internalization and integrity within cultures. In M. R. Leary & J. P. Tangney (Eds.), *Handbook of self and identity* (pp. 253–272). New York: Guilford.

Ryff, C. D., & Keyes, C. L. M. (1995). The structure of psychological well-being revisited. *Journal of Personality and Social Psychology, 69*, 719–727.

Ryff, C. D., & Singer, B. (2003). Flourishing under fire: Resilience as a prototype of challenged thriving. In C. L. M. Keyes & J. Haidt (Eds.), *Flourishing: Positive psychology and the life well-lived* (pp. 15–36). Washington, DC: American Psychological Association.

Sampson, R. J., & Laub, J. H. (1993). *Crime in the making: Pathways and turning points*. Cambridge, MA: Harvard University Press.

Seligman, M. E. P., & Csikszentmihalyi, M. (2000). Positive psychology: An introduction. *American Psychologist, 55*, 5–14.

Spitzer, R. L., Williams, J. B. W., Gibbon, M., & First, M. B. (1990). *Structured Clinical Interview for DSM–III–R.* Washington, DC: American Psychiatric Press.

Taylor, S. E., & Brown, J. D. (1988). Illusion and well-being: A social psychological perspective on mental health. *Psychological Bulletin, 103,* 193–210.

U.S. Department of Health and Human Services. (1999). *Mental health: A report of the Surgeon General.* Rockville, MD: U.S. Department of Health and Human Services, Substance Abuse and Mental Health Services Administration, Center for Mental Health Services, National Institute of Health, National Institute of Mental Health.

U.S. Public Health Service. (1999). *Mental health: A report of the Surgeon General.* Rockville, MD: U.S. Department of Health and Human Services, National Institute of Health, National Institute of Mental Health.

U.S. Public Health Service. (2000). *Report of the Surgeon General's Conference on Children's Mental Health: A national action agenda.* Washington, DC: Department of Health and Human Services.

Vaillant, G. E., & Vaillant, C. O. (1981). Natural history of male psychological health, X: Work as a predictor of positive mental health. *American Journal of Psychiatry, 138,* 1443–1440.

Weissman, M. M. (1995). *Social Adjustment Scale: Reference to publications using the various versions of the scale and information on translations.* Unpublished manuscript, College of Physicians and Surgeons of Columbia University and New York State Psychiatric Institute.

Weissman, M. M., & Paykel, E. S. (1974). *The depressed woman: A study of social relationships.* Chicago: University of Chicago Press.

Werner, E. E., & Smith, R. S. (2001). *Journeys from childhood to midlife: Risk, resilience, and recovery.* Ithaca, NY: Cornell University Press.

White, R. W. (1959). Motivation reconsidered: The concept of competence. *Psychological Review, 66,* 297–333.

RESEARCH IN HUMAN DEVELOPMENT, *1*(4), 327–346

Diversity in Individual ↔ Context Relations as the Basis for Positive Development Across the Life Span: A Developmental Systems Perspective for Theory, Research, and Application

(The 2004 Society for the Study of Human Development Presidential Address)

Richard M. Lerner
Tufts University

Contemporary, cutting-edge scholarship in human developmental science is framed by developmental systems theories emphasizing that the basic process in development involves mutually influential relations between the individually distinct person and his or her diverse, multilevel context; stressing that relative plasticity in development derives from such individual ↔ context relations; and providing optimism about the possibility that applications of developmental science may promote positive development. These ideas underscore the substantive importance of human diversity, seen both as the potential for systematic intraindividual change across life (plasticity) and as interindividual differences in intraindividual change, and stand in contrast to views about diversity as either unimportant, error, or reflecting deficits in development. I discuss these differing conceptions of human diversity and of their distinct implications for the methodological and ethical conduct of science and of applications to programs and policies aimed at promoting positive development.

Eons ago, I began graduate school. It was January 1966, and I was a new doctoral student at the City University of New York. I was told in both of the two required

Requests for reprints should be sent to Richard M. Lerner, Eliot-Pearson Department of Child Development, 105 College Avenue, Tufts University, Medford, MA 02155. E-mail: richard.lerner@tufts.edu substantive courses I took during my first semester—Learning Theory and Experimental Psychology—that there were two fundamental assumptions of all of science. They were that the universe was uniform and that it was permanent. I was further instructed that what these assumptions meant for the study of human behavior was that psychology was a field that sought to identify nomothetic laws that pertained to the generic human being.

From this perspective about science, I learned as well that time and place were irrelevant to the existence of laws of behavior. A law was a valid and hence, from this view, universally generalizable relation among variables involved in an organism's behavior. As such, a law that existed yesterday was a law that existed today and would be a law that existed tomorrow and forever after. A law that existed in New York City would exist as well in Los Angeles; Oaxaca, Mexico; Berlin, Germany; Ann Arbor, Michigan; or on the Moon or Mars for that matter. Indeed, if one found that relations among variables varied across time and place, then this was—by definition—evidence that a basic law was not being identified. What was a basic law, from this perspective? It was a relation among variables that was critical to know to understand what was normative or, in fact, universal about human behavior, or (and perhaps insidiously) to understand what it meant to be human.

As such, individual differences—diversity—were seen, at best, through a lens of error variance, as prima facie proof for instance of a lack of experimental control or of inadequate measurement. At worst, the presence of diversity across time or place or in regard to individual differences among people was regarded as an indication that there was a deficit present. Either the person doing the research was remiss for using a research design or measurement model that was obviously replete with error (i.e., with a lack of experimental control sufficient to eliminate interindividual differences), or the people who varied from the norms associated with the generic human being—the relations among variables that were generalizable across time and place—were in some way deficient (cf. Gould, 1981, 1996). They were, to at least some observers, less than normatively human.

FROM DEFICIT TO DIVERSITY

For those colleagues who were trained in psychology, or more specifically, in one or another area of human development, more recently than I was, say within the last decade or so, the assumptions about science in which I was imbued as a graduate student may seem either unbelievably naïve or simply quaint vestiges from an unenlightened past. Indeed, in what, for the history of science, has been a very short period, participants in the field of human development have seen a major sea change, perhaps one that qualifies as a true paradigm shift in what is regarded as the nature of human nature and of the appreciation of time, place, and individual diversity for understanding the laws of human behavior and development.

For instance, in the 32 years between the beginning of my graduate training and the publication in 1998 of the fifth edition of the *Handbook of Child Psychology*, edited by William Damon, the field of human development has seen a rejection of the hegemony of positivism and reductionism. As evidenced by the chapters in all four volumes of the Damon (1998) *Handbook* and arguably especially in Volume 1, which was entitled "Theoretical Models of Human Development" (Lerner, 1998), the majority of the scholarship that today defines the cutting edge of the field of human development embraces dynamic, developmental systems models of human development. In this view, human development is a product of dynamic or fused relations among levels of organization ranging from inner biology through culture, the natural and designed ecology, and history (Lerner, 2002).

The view of the world that emerges from the chapters in the Damon (1998) *Handbook* is that the universe is dynamic and variegated. Time and place, therefore, are matters of substance, not error; and to understand human development, one must appreciate how variables associated with person, place, and time coalesce to shape the structure and function of behavior and its systematic and successive change (Elder, 1998; Elder, Modell, & Parke, 1993; Magnusson, 1999a, 1999b; Magnusson & Stattin, 1998). Accordingly, diversity of person and context become the foreground of the analysis of human development (Lerner, 1991). From the dynamic, developmental systems perspective framing the contemporary study of human development, there is not a rejection of the idea that there may be general laws of human development. Rather, there is the insistence on the presence of individual laws as well and that any generalizations about groups or humanity as a whole require empirical verification, not preempirical stipulation (Lerner, 2002; Magnusson & Stattin, 1998).

In essence, to paraphrase the insight of Kluckhohn and Murray (1948) made more than a half century ago—and a fitting citation to make here, given the central role of the Murray Research Center in launching and sustaining the Society for the Study of Human Development—all people are like all other people, all people are like some other people, and each person is like no other person. Today, then, the science of human development recognizes that there are idiographic, differential, and nomothetic laws of human behavior and development (e.g., see Emmerich, 1968; Lerner, 2002) and that each person and each group possesses unique and shared characteristics that need to be the core targets of developmental analysis.

The Persistence of Reductionist Models

Unfortunately, however, despite the contemporary emphasis on developmental systems models, the remnants of reductionism and deficit thinking still remain within the field of human development. For instance, they exist in the behavior genetics (e.g., Plomin, 2000; Rowe, 1994), sociobiology (e.g., Rushton, 1999,

2000), and (in at least some forms of the) evolutionary psychology (e.g., Buss, 2003) instantiations of genetic reductionism. These approaches constitute today's version of the biologizing errors of the past such as eugenics and racial hygiene (Proctor, 1988). As explained by Collins, Maccoby, Steinberg, Hetherington, and Bornstein (2000), these ideas are no longer seen as part of the forefront of scientific theory. Nevertheless, their influence on scientific and public policy persist. Indeed, renowned geneticists, such as Bearer (2004), Edelman (1987, 1988), Feldman (e.g., Feldman & Laland, 1996), Ho (1984), Lewontin (2000), Müller-Hill (1988), and Venter (e.g., Venter et al., 2001), and eminent colleagues in comparative and biological psychology, such as Greenberg (e.g., Greenberg & Haraway, 2002; Greenberg & Tobach, 1984), Gottlieb (1997, 2004), Hirsch (1997, 2004), Michel (e.g., Michel & Moore, 1995), and Tobach (1981, 1994; Tobach, Gianutsos, Topoff, & Gross, 1974), alert us to the need for continued intellectual and social vigilance lest such flawed ideas about genes and human development become the foci of public policies or social programs.

Such unfortunate applications of counterfactual ideas remain real possibilities and in some cases, unfortunate realities due at least in part to what Horowitz (2000) described as the affinity of the "Person in the Street" to simplistic models of genetic affects on behavior. These simplistic and, I must emphasize, erroneous models are used by the Person in the Street to form opinions or to make decisions about the nature of human differences and potentials.

In essence, genetic reductionism can and has led to views of diversity as a matter of the haves and the have-nots (e.g., Herrnstein & Murray, 1994; Rushton, 1999, 2000). There are, in this view, those people who manifest the normative characteristics of human behavior and development. Given the diversity-insensitive assumptions and research that characterized much of the history of scholarship in human development, even into the 1990s, these normative features of human development were associated with middle-class, Euro-American samples (Graham, 1992; McLoyd, 1998; Spencer, 1990). In turn, there are those people who manifest other characteristics, and these individuals were generally non-Euro-American and non–middle class. However, if the former group is regarded as normative, then the characteristics of the latter groups are regarded as non-normative (Gould, 1996). When such an interpretation is forwarded, entry has thus been made down the slippery slope of moving from a description of between-group differences to an attribution of deficits in the latter groups (Lerner, 2002).

Such an attribution is buttressed when seen through the lens of genetic reductionism because it must be genes in this conception that provide the final, material, and efficient cause of the characteristics of the latter groups (e.g., see Rowe, 1994; Rushton, 2000). These non-Euro-American and/or non-middle-class groups are, in the fully tautological reasoning associated with genetic reductionism, behaviorally deficient because of the genes they possess and because of the genes they possess they have behavioral deficits (e.g., see Rushton, 2000).

Simply, the genes that place one in a racial group are the genes that provide either deficit or assets in behavior, and one racial group possesses the genes that are assets, and the other group possesses the genes that are deficits.

As shown in Table 1, these genetic reductionist ideas may have profound and dire affects on public policies and social programs (Lerner, 2004). Table 1 presents "A.," beliefs about whether genetic reductionist ideas are believed to be either (1) true or (2) false. The table presents also "B.," public policy and social program implications that would be associated with genetic reductionism were it in fact (1) true or (2) false under either of the two belief conditions involved in A. Moreover, the A.2.B.2. quadrant of the table not only presents the policy and pro-

TABLE 1
Policy and Program Implications That Arise If the Genetic Reductionist, "Split" Conception of Genes (A) Were Believed to Be True or False and (B) Were in Fact True or False

		B. Public Policy and Social Program Implications If Genetic Reductionist Position Were in Fact	
		1. True	*2. False*
A. Genetic Reductionist Conception Is Believed to Be	1. True	• Repair inferior genotypes, making them equal to superior genotypes • Miscegenation laws • Restrictions of personal liberties of carriers of inferior genotypes (separation, discrimination, distinct social tracts) • Sterilization • Elimination of inferior genotypes from genetic pool	• Same as A1, B1
	2. False	• Wasteful and futile humanitarian policies • Wasteful and futile programs of equal opportunity, affirmative action, equity, and social justice • Policies and programs to quell social unrest because of unrequited aspirations of genetically constrained people • Deterioration of culture and destruction of civil society	• Equity, social justice, equal opportunity, affirmative action • Celebration of diversity • Universal participation in civic life • Democracy • Systems assessment and engagement • Civil society

gram implications of believing that the genetic reductionist conception is believed to be false when it is in fact false. In addition, this quadrant illustrates the policy and program implications of believing developmental systems theory to be true when it is in fact the case, as I obviously contend, that it is true. Table 1 demonstrates that if genetic reductionism is believed to be true, then irrespective of whether it is in fact true (and it must be emphasized that it is incontrovertibly not true), a range of actions may be promoted that constrain people's freedom of association, reproductive rights, and even survival.

Examples of Reductionist and Deficit Thinking in Contemporary Public Discourse

One may think that a straw man is being attacked here and that in today's world, neither the Person in the Street nor leading instances of the media informing this person would promote the egregiously flawed and counterfactual ideas associated with genetic reductionism. I briefly describe, then, three recent stories appearing in *The Washington Post* as evidence that genetic reductionism and in effect therefore, the deficit model of human development, are still very much a part of public discourse.

An August 1, 2003 front page story by *The Washington Post* staff writer Shanker Vedantam appeared under the headline "Desire and DNA: Is Promiscuity Innate? New Study Sharpens Debate on Men, Sex, and Gender Roles." Vedantam's (2003) article reported the findings of a study of "more than 16,000 people from every inhabited continent [that] found that men everywhere—whether single, married or gay—want more sexual partners than women do" (p. A1). The study was published in the *Journal of Personality and Social Psychology* by David P. Schmitt (2003) of Bradley University. Schmitt, a self-described evolutionary psychologist, reported the results of a survey given to volunteer adults from more than 50 nations. Despite the fact that the measurement model of this research was limited to only verbal reports assessed among literate people willing and able to participate in survey research, neither limitations of the sample nor of the measurement model kept Schmitt from noting that "The results are strong and conclusive—the sexes differ, and these differences appear to be universal" (Vedantam, 2003, p. A1). Similarly, the limitations of this research did not deter David Buss (e.g., 2003), another evolutionary psychologist from the University of Texas at Austin, from asserting that "The evidence he [Schmitt] presents is irrefutable" (Vedantam, 2003, p. A8).

Consider as well the comparably irrefutable evidence that columnist George Will described in the Sunday, September 21, 2003 edition of *The Washington Post*. I quote here the first paragraph of the column:

> Science is reshaping the argument about whether nature or nurture is decisive in determining human destinies and about what the answer means for social policy. Consider a fascinating new report arguing the scientific evidence for the importance of "authoritative communities"—groups, religious or secular, devoted to transmitting a model of the moral life. The report is from 33 research scientists, children's doctors and mental health and youth services professionals on a commission jointly sponsored by the Dartmouth Medical School, the Institute for American Values and the YMCA of the USA. The report's conclusion is in its title: Human beings are "Hardwired to Connect." (Will, 2003, p. B7)

Although many of the authors of this report (Commission on Children at Risk, 2003) subscribe to developmental systems models of behavior and development, their unfortunate use of the phrase "hard wired" in the title of the report was sufficient for Will (2003) to claim that the problem facing some youth of the nation was—and again I quote

> A deficit of connectedness. The deficit is the difference between what the biological makeup of human beings demands and what many children's social situations supply in the way of connections to other people and to institutions that satisfy the natural need for moral and spiritual meaning. (p. B7)

However, at the end of the Will (2003) column and after claiming that "there may be a biological basis for religious affiliation" (p. B7), Will noted that the report "suggests that there is no simple 'versus' in nature versus nurture" (p. B7) and that "There is a complex interaction." Nevertheless, these qualifications of the scientific message in the report are too little and too late in Will's presentation to enlighten most readers about the subtleties of interrelations within the dynamic developmental system. I think that the key idea about nature and nurture that the Person in the Street is likely to derive from this column is that genes provide the basis for morality, spirituality, and religious affiliation and that the role of the context is to facilitate or—in deficient families or communities—frustrate the unfolding of these primarily intrinsic attributes.

Thus, even when reporting results of research framed by or consistent with the developmental systems perspective, there is a pronounced tendency to give legitimacy to the reductionist view of genetic action and of the role of context as independent of or split from genes (see Overton, 1998). For instance, in the September 2 issue of *The Washington Post,* staff writer Rick Weiss (2003) reported a story under the headline "Genes' Sway Over IQ May Vary With Class." Describing the developmental systems-oriented work of University of Virginia psychologist Eric Turkheimer (Turkheimer, Haley, Waldron, D'Onofrio, & Gottesman, 2003), published in *Psychological Science*, Weiss noted that the study of the interaction among genes, environment, and IQ indicates that the influence of genes on intelligence is dependent on social class. Weiss described Turkheimer et al.'s findings as indicating that genes explain the intelligence scores of middle-class and upper class children, wherein the heritability of IQ is estimated to be about 0.7, but not of lower socioeconomic children, wherein heritability is about 0.1.

Weiss (2003) pointed to Turkheimer et al.'s (2003) work as representative of a dynamic view of gene–environment relations, one that involves both the influence of genes on the impact of experiences and of the impact of experience on gene expression (i.e., that a gene ↔ experience relation, or better, coaction exists). This view, Weiss noted, transcends what he depicts as the older "heritability" tradition of viewing nature and nurture as largely independent and additive factors. Nevertheless, Weiss then immediately used this older tradition to note that genes do explain the vast majority of IQ differences among children in wealthy families. Indeed, in what seems like a clear attempt to present a fair and balanced report of the differences between two equally viable positions—the newer, dynamic one versus the older, heritability view—Weiss drew on comments from Sandra Scarr, Irving Gottesman, and Robert Plomin as representatives of the latter position. Scarr's comments pertain to the usefulness of twin or adoption designs to study the heritability of intelligence, and Gottesman's comments stress the point that high heritability does not mean that interventions cannot be effective.

However, Plomin, who is described as a researcher working to identify the genes linked specifically to intelligence, is cited as maintaining that Turkheimer et al.'s (2003) results "do not undermine the importance of genes" (Weiss, 2003). Plomin is quoted as saying that "In study after study, the evidence is overwhelming that there is a substantial genetic input to IQ. This [Turkheimer's work] doesn't contradict that."

In short, despite the theory and research that lends support to a dynamic conception of gene ↔ experience coaction, there exist proponents of genetic reductionism who maintain that concepts and methods that regard genes as separable from context are valid and overwhelmingly or irrefutably evident. The media continue to tell this story, and perhaps more often than not, the Person in the Street is persuaded by it.

The challenge that such language use and public discourse represents is not merely one of meeting developmental scientists' scientific responsibility to act to amend incorrect dissemination of research evidence. Horowitz (2000) reminded developmental scientists that an additional, and ethical, responsibility that we face is to act in a manner supportive of social justice. Indeed, Horowitz emphasized that such action is critical in the face of the simplistically seductive ideas of genetic reductionism, especially when coupled with the deficit model. Horowitz (2000) explained that

> If we accept as a challenge the need to act with social responsibility then we must make sure that we do not use single-variable words like genes or the notion of innate

in such a determinative manner as to give the impression that they constitute the simple answers to the simple questions asked by the Person in the Street lest we contribute to belief systems that will inform social policies that seek to limit experience and opportunity and, ultimately, development, especially when compounded by racism and poorly advantaged circumstances. Or, as Elman and Bates and their colleagues said in the concluding section of their book *Rethinking Innateness* (Elman et al., 1998), "If our careless, under-specified choice of words inadvertently does damage to future generations of children, we cannot turn with innocent outrage to the judge and say 'But your Honor, I didn't realize the word was loaded.' " (p. 8)

THE DEVELOPMENTAL SYSTEMS PERSPECTIVE

There exists another vision of and vocabulary for the role of genes and of biology more generally in human development. As illustrated in Table 1, the dynamic developmental systems models of human development that are today at the fore of contemporary scholarship in human development (Lerner, 1998, 2002) and that provide a much different view of the role of genes in behavior and development offer a much different, if admittedly more complex, story to the Person in the Street. These developmental systems models emphasize that human development is a relational phenomenon (Lerner, 1991, 2002; Overton, 1998). Eschewing splits between nature and nurture, organism and environment, or any of the other Cartesian dualities that have been part of the discourse in modern developmental science (e.g., continuity–discontinuity, maturation–experience, or stability–instability; see Overton, 1998), developmental systems theories stress that genes, cells, tissues, organs, whole organisms, and all other, extraorganism levels of organization comprising the ecology of human development are fused in a fully coactional, mutually influential, and therefore dynamic system (Bronfenbrenner, 2001, 2004; Gottlieb, 1997, 1998, 2004; Tobach, 1981).

This bidirectional relation between the individual and the complex ecology of human development may be represented as individual ↔ context. The fact that the broadest level of the context is history means that temporality is always a part of the fused system of individual ↔ context relations and thus that the potential for systematic change, that is, for plasticity, exists across the life span (Baltes, Lindenberger, & Staudinger, 1998; Elder, 1998). Of course, the system that promotes change through the coaction of multiple levels of organization can also act to constrain it. Therefore, this fusion of the potential for both constancy and change makes plasticity relative and not absolute (Lerner, 1984).

Nevertheless, the temporality of human development and the presence of at least relative plasticity indicate that one may be optimistic that means may be found at one or more levels of the ecology of human development to apply developmental science in ways that promote positive development across the life span (Bronfenbrenner, 2004; Ford & Lerner, 1992; Lerner, 2002, 2004; Magnusson & Stattin, 1998). Moreover, because no two people, even monozygotic (MZ) twins, will have the same history of individual ↔ context relations across the life span, the individuality of each person is lawfully assured (Hirsch, 1970, 1997, 2004). Indeed, given that there are over 70 trillion potential human genotypes, that the probability of two genetically identical children arising from any set of parents is 1 in 6.27 billion, and that the probability of two genetically identical but non-MZ children arising from one specific couple is slightly less than 1 in 160,000 (Hirsch, 2004), there is an obviously low probability that any two people, with the exceptions of MZs, will have an identical biological genotype (to use a redundancy).

However, the probability that two different people, including MZs, will have a history of events, experiences, and social relationships, that is, a social genotype (to use an oxymoron), that is identical is so dismally small as to be equivalent to what most people would regard as impossible. In other words, the integration of biology and context across time means that each person has a developmental trajectory (a dynamically changing phenotype) that is at least in part individually distinct.

Diversity Is a Fundamental Asset of Human Development

Diversity is, then, a distinctive and in fact, a defining feature of the human life course. Within an individual over time, diversity, seen as the potential for systematic intraindividual change, represents a potential for life-span change. Therefore, diversity as intraindividual plasticity is a key "asset," or developmental strength, that may be capitalized on to promote a person's positive, healthy developmental change. Across people, diversity—interindividual differences—represents a sample of the range of variation that defines the potential material basis for optimizing the course of human life. In other words, any individual may have a diverse range of potential developmental trajectories and as well, all groups—because of the necessarily diverse developmental paths of the people within it—will have a diverse range of developmental trajectories. In short, diversity, seen as both intraindividual change and as interindividual differences in intraindividual change, is both a strength of individuals and an asset for planning and promoting means to improve the human condition.

The diversity of individual ↔ context relations that comprises change within the dynamic developmental system and the optimism about improving human life that derives from the relative plasticity of humans means that it is possible to apply developmental science to promote positive development across the life span (Lerner, 2002, 2004). In the remainder of this article, I describe briefly the features and implications for science and application of the positive human development perspective derived from developmental systems theories.

FEATURES AND IMPLICATIONS OF A POSITIVE HUMAN DEVELOPMENT PERSPECTIVE

The fused system of individual ↔ context relations that provides each person with the potential for relative plasticity across the life span constitutes a fundamental strength of each person. This strength is present to differing extents in all infants, children, adolescents, adults, and aged individuals. Relative plasticity diminishes across the life span, but as the research of Baltes in the Berlin Study of Aging (e.g., Baltes et al., 1998; Baltes & Smith, 2003; Baltes, Staudinger, & Lindenberger, 1999; Smith et al., 2002) elegantly demonstrates, there is evidence for the presence of plasticity into the 10th and 11th decades of life.

The fused developmental system provides also a potential for change not just in people but as well in the contexts within which individuals develop. This latter potential means that families, neighborhoods, and cultures are also relatively plastic and that the level of resources—or developmental assets—that they possess at any point in time may also be altered across history. Contextual strengths and assets in support of positive development may be envisioned within the terms suggested by Benson (2003), as the community "nutrients" for healthy and positive development. These assets can be grown, aligned, and realigned to improve the circumstances of human development.

Of course, at any given place or point in time, both individuals and levels of the context within this plastic developmental system may manifest problems or may be deficient in some aspect of individual, family, or community life that is needed for improved functioning. Obviously, the presence of plasticity does not mean that people are not poor or that they do not lack some social nutrients that would enhance their development. However, what the presence of relative plasticity across the developmental system does mean is that all individuals have strengths that when integrated with the developmental assets of communities may be capitalized on to promote positive change. As such, problems or deficits constitute only a portion of a potentially much larger array of outcomes of individual ↔ context relations. Problems are not inevitable, and they are certainly not fixed in one's genes.

The role of developmental science is to identify those relations between individual strengths and contextual assets in families, communities, cultures, and the natural environment and to integrate strengths and assets to promote positive human development (Lerner, 2004). A system that is open for change for the better is also open for change for the worse. The research and applications of developmental scientists should be aimed at increasing the probability of actualization of the healthy and positive portions of the distribution of potential outcomes of individual ↔ context relations.

In other words, the scientific agenda of the developmental scientist is more than just to describe and to explain human development. It is also to work to optimize it (Baltes, 1968). Efforts to enhance human development in its actual ecology are a way to test one's theoretical ideas about how systemic relations coalesce to shape the course of life. These efforts stand as well as ethical responsibilities of human development scholars in regard to both their roles as researchers involved with human lives and as citizens of a civil society (Fisher, 1993, 1994, 2003; Fisher, Hoagwood, & Jenson, 1996; Fisher & Tryon, 1990).

Moreover, without a scientific agenda that integrates description, explanation, and optimization, human development science is at best an incomplete scholarly endeavor. That is, a developmental science that is devoid of knowledge of the individual and group ranges among diverse groups, and a developmental science that is devoid of knowledge of the range of assets in diverse contexts, is an incomplete developmental science. It is also an inadequate developmental science when seen from the perspective of the need for evidence-based policy and program applications.

Framing the Research Agenda of Human Development

What becomes, then, the key empirical question for developmental scientists interested in describing, explaining, and promoting positive human development? The key question is actually five interrelated "what" questions:

1. what attributes?; of
2. what individuals?; in relation to
3. what contextual/ecological conditions?; at
4. what points in ontogenetic, family or generational, and cohort or historical, time?; may be integrated to promote
5. what instances of positive human development?

Answering these questions requires a nonreductionist approach to methodology. Neither biogenic, psychogenic, nor sociogenic approaches are adequate. Developmental science needs integrative and relational models, measures, and designs (Lerner, Dowling, & Chaudhuri, in press). Examples of the use of such methodology within developmental systems oriented research are the scholarship of Eccles and her colleagues (e.g., Eccles, Wigfield, & Byrnes, 2003) on stage ↔ environment fit; of Damon (1997; Damon & Gregory, 2003) on the community-based youth charter; of Benson and his colleagues (e.g., Benson, Leffert, Scales, & Blyth, 1998; Leffert et al., 1998; Scales, Benson, Leffert, & Blyth, 2000) at Search Institute on the role of developmental assets in positive youth development; and of Leventhal and Brooks-Gunn (2004) and of Sampson, Raudenbush, and Earls (1997) on the role of neighborhood characteristics on adolescent development.

Another example of integrative, relational developmental research may be found in the 4-H study of positive youth development, which I am conducting with my colleague, Jacqueline V. Lerner, and our colleagues and students at Tufts University and at Boston College, respectively (Lerner, Lerner, et al., in press). Within the burgeoning theoretical ideas associated with the concept of positive youth development (PYD) perspective (Benson, 2003; Lerner, 2004), exemplary adolescent development has been theorized to involve competence, confidence, caring, character, and positive social connections. These "five Cs" are believed to be derived from developmental assets and perhaps especially from participation in exemplary community youth development programs (e.g., Eccles & Gootman, 2002; Roth & Brooks-Gunn, 2003a, 2003b). In addition, theory and the data converge in suggesting that when the five Cs are present in a young person's development, there is the emergence of a superordinate "sixth C": contribution to self, family, community, and civil society (Lerner, 2004; Lerner, Brentano, Dowling, & Anderson, 2002; Lerner, Fisher, & Weinberg, 2000).

The 4-H study (Lerner, Lerner, et al., in press) of PYD is the first comprehensive longitudinal data set exploring the empirical composition of the five Cs, their convergence into an overall construct of PYD, and the purported role of developmental assets in general, of youth development program participation more specifically, and of 4-H participation in particular in mediating the development of the five Cs and of the sixth C of community contribution. Structural equation modeling (SEM; i.e., LISREL, Version 8.53; Jöreskog & Sörbom, 2002) results from the first wave of data collection of the 4-H study—which involved approximately 1,700 fifth graders who varied in race, ethnicity, socioeconomic status, family structure, rural–urban location, geographic region, and program participation experiences and who were sampled from schools in 13 states from across the nation—provide the first evidence to date of the empirical reality of the five Cs (Lerner, Lerner, et al., in press). The SEM findings indicated strong evidence for first-order latent variables representing the Cs of PYD and for their convergence on a second-order latent construct of PYD itself. In turn, youth involved in 4-H programs did not differ from other youth in regard to mean scores for any of the five Cs. However, the 4-H youth, independent of (i.e., over and above) their participation in 4-H community activities, had significantly higher scores for community contribution. Moreover, frequency of community participation was significantly predicted by 4-H program participation (Lerner, Lerner, et al., in press).

As illustrated also by the 4-H study (Lerner, Lerner, et al., in press), the methodology employed in individual ↔ context integrative research must also include a triangulation among multiple and, ideally, both qualitative and quantitative approaches to understanding and synthesizing variables from the levels of organization within the developmental system. Such triangulation may usefully involve the "classic" approach offered by Campbell and Fiske (1959) regarding convergent and discriminant validation through multitrait-multimethod matrix methodology. Of course, diversity-sensitive measures are needed within such approaches, and they must be used within the context of change-sensitive—and hence longitudinal—designs (Lerner, Dowling, et al., in press; Magnusson & Stattin, 1998). Trait measures developed with the goal of excluding variance associated with time and context are clearly not optimal choices in such research. In other words, to reflect the richness and strengths of a diverse humanity, developmental scientists' repertoire of measures must be sensitive to the diversity of person variables, such as race, ethnicity, religion, sexual preferences, physical ability status, and developmental status, and to the diversity of contextual variables such as family type, neighborhood, community, culture, physical ecology, and historical moment.

Indeed, it is particularly important that designs and measures must also be sensitive to the different meanings of time. Insightful formulations about the different meaning of time within the dynamic developmental system have been provided by Elder (1998), Baltes et al. (1998, 1999), and Bronfenbrenner (2004). Methods must appraise, then, age, family, and historical time and must be sensitive to the role of both normative and non-normative historical events in influencing developmental trajectories.

Finally, developmental scientists' designs should be informed by not only colleagues from the multiple disciplines with expertise in the scholarly study of human development. The methods should be informed as well by the individuals and communities we study (Lerner, 2002, 2004; Villarruel, Perkins, Borden, & Keith, 2003). They too are experts about development, a point our colleagues in cultural anthropology, sociology, and community youth development research and practice have been making for several years. Most certainly, participants in our community-based research and applications are experts in regard to the character of development within their families and neighborhoods. Accordingly, research that fails to capitalize on the wisdom of its participants runs the real danger of lacking authenticity and of erecting unnecessary obstacles to the translation of the scholarship of knowledge generation into the scholarship of knowledge application (Jensen, Hoagwood, & Trickett, 1999).

CONCLUSIONS

Is the approach that I have discussed to the description, explanation, and application of developmental science in the service of promoting positive human development more complex than the reductionist formulations that often attract the Person in the Street? Yes. However, it is also a richer and ecologically valid one. It is an approach to developmental science that underscores the diverse ways in which humans, in dynamic exchanges with their natural and designed ecologies, can create for themselves and others opportunities for health and positive development. As Bronfenbrenner (2005) eloquently put it, it is these relations that make human beings human.

The relational, dynamic, and diversity-sensitive developmental science that now defines the cutting edge of our field—a cutting edge shaped in major ways by colleagues in the Society for the Study of Human Development—may both document and extend the power inherent in each person to be an active agent in his or her own successful and positive development (Brandtstädter, 1998, 1999; Lerner, 1982; Lerner & Busch-Rossnagel, 1981; Lerner, Theokas, & Jelicic, in press; Lerner & Walls, 1999). By celebrating the strengths of all individuals and the assets that exist in their families, communities, and cultures to promote these strengths, we can have a developmental science that may, in these challenging times, help us, as a scientific body and as citizens of a democratic nation, finally ensure that there is truly liberty and justice for all.

ACKNOWLEDGMENTS

This article is based on the Presidential Address presented at the Third Biennial Meeting of the Society for the Study of Human Development. I thank my colleagues and students at the Institute for Applied Research in Youth Development in the Eliot-Pearson Department of Child Development for their comments on prior versions of this article. The preparation of this article was supported in part by grants from the National 4-H Council and from the William T. Grant Foundation.

REFERENCES

Baltes, P. B. (1968). Longitudinal and cross-sectional sequences in the study of age and generation effects. *Human Development, 11*, 145–171.

Baltes, P. B., Lindenberger, U., & Staudinger, U. M. (1998). Life-span theory in developmental psychology. In W. Damon (Series Ed.) & R. M. Lerner (Vol. Ed.), *Handbook of child psychology: Vol. 1. Theoretical models of human development* (5th ed., pp. 1029–1144). New York: Wiley.

Baltes, P. B., & Smith, J. (2003). New frontiers in the future of aging: From successful aging of the young old to the dilemmas of the fourth age. *Gerontology, 49*, 12–135.

Baltes, P. B., Staudinger, U. M., & Lindenberger, U. (1999). Lifespan psychology: Theory and application to intellectual functioning. In J. T. Spence, J. M. Darley, & D. J. Foss (Eds.), *Annual review of psychology* (Vol. 50, pp. 471–507). Palo Alto, CA: Annual Reviews.

Bearer, E. (2004). Behavior as influence and result of the genetic program: Non-kin rejection, ethnic conflict and issues in global health care. In C. Garcia Coll, E. Bearer, & R. M. Lerner (Eds.), *Nature and nurture: The complex interplay of genetic and environmental influences on human behavior and development* (pp. 171–199). Mahwah, NJ: Lawrence Erlbaum Associates, Inc.

Benson, P. L. (2003). Developmental assets and asset-building community: Conceptual and empirical foundations. In R. M. Lerner & P. L. Benson (Eds.) *Developmental assets and asset-building com-*

munities: Implications for research, policy, and practice (pp. 19–43). Norwell, MA: Kluwer Academic.

Benson, P. L., Leffert, N., Scales, P. C., & Blyth, D. A. (1998). Beyond the "village" rhetoric: Creating healthy communities for children and adolescents. *Applied Developmental Science, 2,* 138–159.

Brandtstädter, J. (1998). Action perspectives on human development. In W. Damon (Series Ed.) & R. M. Lerner (Vol. Ed.), *Handbook of child psychology: Vol. 1. Theoretical models of human development* (5th ed., pp. 807–863). New York: Wiley.

Brandtstädter, J. (1999). The self in action and development: Cultural, biosocial, and ontogenetic bases of intentional self-development. In J. Brandtstädter & R. M. Lerner (Eds.), *Action and self-development: Theory and research through the life-span (pp. 37–65). Thousand Oaks, CA: Sage.*

Bronfenbrenner, U. (2001). Human development, bioecological theory of. In N. J. Smelser & P. B. Baltes (Eds.), *International encyclopedia of the social and behavioral sciences* (pp. 6963–6970). Oxford, England: Elsevier.

Bronfenbrenner, U. (2005). *Making human beings human.* Thousand Oaks, CA: Sage.

Buss, D. M. (2003). *Evolutionary psychology: The new science of the mind* (2nd ed.). Boston: Allyn & Bacon.

Campbell, D. T., & Fiske, D. W. (1959). Convergent and discriminant validation by the multitrait-multimethod matrix. *Psychological Bulletin, 56,* 81–105.

Collins, W. A., Maccoby, E. E., Steinberg, L., Hetherington, E. M., & Bornstein, M. H. (2000). Contemporary research on parenting: The case of nature and nurture. *American Psychologist, 55,* 218–232.

Commission on Children at Risk. (2003). *Hardwired to connect: The new case for authoritative communities.* New York: YMCA of the USA, Dartmouth Medical School, Institute for American Values.

Damon, W. (1997). *The youth charter: How communities can work together to raise standards for all our children.* New York: Free Press.

Damon, W. (Ed.). (1998). *Handbook of child psychology* (5th ed.). New York: Wiley.

Damon, W., & Gregory, A. (2003). Bringing in a new era in the field of youth development. In R. M. Lerner, F. Jacobs, & D. Wertlieb (Eds.), *Applying developmental science for youth and families: Historical and theoretical foundations.* Volume 1 of *Handbook of applied developmental science: Promoting positive child, adolescent, and family development through research, policies, and programs* (pp. 407–420). Thousand Oaks, CA: Sage.

Eccles, J., & Gootman, J. A. (Eds). (2002). *Community programs to promote youth development.* Washington, DC: National Academy Press.

Eccles, J., Wigfield, A., & Byrnes, J. (2003). Cognitive development in adolescence. In I. B. Weiner (Series Ed.) & R. M. Lerner, M. A. Easterbrooks, & J. Mistry (Vol. Eds.), *Handbook of psychology: Vol. 6. Developmental psychology* (pp. 325–350). New York: Wiley.

Edelman, G. M. (1987). *Neural Darwinism: The theory of neuronal group selection.* New York: Basic Books.

Edelman, G. M. (1988). *Topobiology: An introduction to molecular biology.* New York: Basic Books.

Elder, G. H., Jr. (1998). The life course and human development. In W. Damon (Series Ed.) & R. M. Lerner (Vol. Ed.), *Handbook of child psychology: Vol. 1. Theoretical models of human development* (5th ed., pp. 939–991). New York: Wiley.

Elder, G. H., Modell, J., & Parke, R. D. (1993). Studying children in a changing world. In G. H. Elder, J. Modell, & R. D. Parke (Eds.), *Children in time and place: Developmental and historical insights* (pp. 3–21). New York: Cambridge University Press.

Elman, J. L., Bates, E. A., Johnson, M. H., Karmiloff-Smith, A., Parisi, D., & Plunkett, K. (1998). *Rethinking innateness: A connectionist perspective on development (neural network modeling and connectionism).* Cambridge, MA: MIT Press.

Emmerich, W. (1968). Personality development and concepts of structure. *Child Development, 39,* 671–690.

Feldman, M. W., & Laland, K. N. (1996). Gene-culture coevolutionary theory. *Trends in Ecology and Evolution, 11*, 453–457.

Fisher, C. B. (1993). Integrating science and ethics in research with high-risk children and youth. *SRCD Social Policy Report, 7,* 1–27.

Fisher, C. B. (1994). Reporting and referring research participants: Ethical challenges for investigators studying children and youth. *Ethics and Behavior, 4*, 87–95.

Fisher, C. B. (2003). *Decoding the ethics code: A practical guide for psychologists*. Thousand Oaks, CA: Sage.

Fisher, C. B., Hoagwood, K., & Jensen, P. (1996). Casebook on ethical issues in research with children and adolescents with mental disorders. In K. Hoagwood, P. Jensen, & C. B. Fisher (Eds.), *Ethical issues in research with children and adolescents with mental disorders* (pp. 135–238). Mahwah, NJ: Lawrence Erlbaum Associates, Inc.

Fisher, C. B., & Tryon, W. W. (1990). Emerging ethical issues in an emerging field. In C. B. Fisher & W. W. Tryon (Eds.), *Ethics in applied developmental psychology: Emerging issues in an emerging field* (pp. 1–15). Norwood, NJ: Ablex.

Ford, D. H., & Lerner, R. M. (1992). *Developmental systems theory: An integrative approach.* Newbury Park, CA: Sage.

Gottlieb, G. (1997). *Synthesizing nature-nurture: Prenatal roots of instinctive behavior*. Mahwah, NJ: Lawrence Erlbaum Associates, Inc.

Gottlieb, G. (1998). Normally occurring environmental and behavioral influences on gene activity: From central dogma to probabilistic epigenesis. *Psychological Review, 105*, 792–802.

Gottlieb, G. (2004). Normally occurring environmental and behavioral influences on gene activity. In C. Garcia Coll, E. Bearer, & R. M. Lerner (Eds.), *Nature and nurture: The complex interplay of genetic and environmental influences on human behavior and development* (pp. 85–106). Mahwah, NJ: Lawrence Erlbaum Associates, Inc.

Gould, S. J. (1981). *The mismeasure of man*. New York: Norton.

Gould, S. J. (1996). *The mismeasure of man* (Rev. ed.). New York: Norton.

Graham, S. (1992). "Most of the subjects were white and middle class": Trends in published research on African Americans in selected APA journals, 1970–1989. *American Psychologist, 47*, 629–639.

Greenberg, G., & & Haraway, M. M. (2002). *Principles of comparative psychology*. Boston: Allyn & Bacon.

Greenberg, G. & Tobach, E. (Eds.). (1984). *Behavioral evolution and integrative levels*. Hillsdale, NJ: Lawrence Erlbaum Associates, Inc.

Herrnstein, R. J., & Murray, C. (1994). *The bell curve: Intelligence and class structure in American life*. New York: Free Press.

Hirsch, J. (1970). Behavior-genetic analysis and its biosocial consequences. *Seminars in Psychiatry, 2*, 89–105.

Hirsch, J. (1997). Some history of heredity-vs-environment, genetic inferiority at Harvard (?), and the (incredible) bell curve. *Genetica, 99,* 207–224.

Hirsch, J. (2004). Uniqueness, diversity, similarity, repeatability, and heritability. In C. Garcia Coll, E. Bearer, & R. M. Lerner (Eds.), *Nature and nurture: The complex interplay of genetic and environmental influences on human behavior and development* (pp. 127–138). Mahwah, NJ: Lawrence Erlbaum Associates, Inc.

Ho, M.-W. (1984). Environment and heredity in development and evolution. In M.-W. Ho & P. T. Saunders (Eds.), *Beyond neo-Darwinism: An introduction to the new evolutionary paradigm* (pp. 267–289). London: Academic.

Horowitz, F. D. (2000). Child development and the PITS: Simple questions, complex answers, and developmental theory. *Child Development, 71*, 1–10.

Jensen, P., Hoagwood, K., & Trickett, E. (1999). Ivory towers or earthen trenches?: Community collaborations to foster real world research. *Applied Developmental Science, 3*, 206–212.

Jöreskog, K. G., & Sörbom, D. (2002). LISREL (Version 8.53) [Computer software]. Lincolnwood, IL: Scientific Software International, Inc.

Kluckhohn, C., & Murray, H. (1948). Personality formation: The determinants. In C. Kluckhohn & H. Murray (Eds.), *Personality in nature, society, and culture* (pp. 35–48). New York: Knopf.

Leffert, N., Benson, P., Scales, P., Sharma, A., Drake, D., & Blyth, D. (1998). Developmental assets: Measurement and prediction of risk behaviors among adolescents. *Applied Developmental Science, 2*, 209–230.

Lerner, R. M. (1982). Children and adolescents as producers of their own development. *Developmental Review, 2*, 342–370.

Lerner, R. M. (1984). *On the nature of human plasticity*. New York: Cambridge University Press.

Lerner, R. M. (1991). Changing organism-context relations as the basic process of development: A developmental contextual perspective. *Developmental Psychology, 27*, 27–32.

Lerner, R. M. (Ed.). (1998). *Handbook of child psychology: Vol. 1. Theoretical models of human development.* (5th ed., Series Ed. W. Damon). New York: Wiley.

Lerner, R. M. (2002). *Concepts and theories of human development* (3rd ed.). Mahwah, NJ: Lawrence Erlbaum Associates, Inc.

Lerner, R. M. (2004). *Liberty: Thriving and civic engagement among America's youth*. Thousand Oaks, CA: Sage.

Lerner, R. M., Brentano, C., Dowling, E. M., & Anderson, P. M. (2002). Positive youth development: Thriving as a basis of personhood and civil society. In R. M. Lerner, C. S. Taylor, & A. von Eye (Vol. Eds.) & G. Noam (Series Ed.), *New directions for youth development: Theory, practice and research: Pathways to positive development among diverse youth* (Vol. 95, pp. 11–34). San Francisco: Jossey-Bass.

Lerner, R. M., & Busch-Rossnagel, N. A. (Eds.). (1981). *Individuals as producers of their development: A life-span perspective*. New York: Academic.

Lerner, R. M., Dowling, E., & Chaudhuri, J. (in press). Methods of contextual assessment and assessing contextual methods: A developmental contextual perspective. In D. M. Teti (Ed.), *Handbook of research methods in developmental psychology*. Cambridge, MA: Blackwell.

Lerner, R. M., Fisher, C. B., & Weinberg, R. A. (2000). Toward a science for and of the people: Promoting civil society through the application of developmental science. *Child Development, 71*, 11–20.

Lerner, R. M., Lerner, J. V., Theokas, C., Jelecic, H., Gestsdottir, S., Alberts, A., Ma, L., Christiansen, E., Almerigi, J., Warren, D., Naudeau, S., Simpson, I., Smith, L. M., & Bentley, A. (in press). Towards a new vision and vocabulary about adolescence: Theoretical and empirical bases of a "positive youth development" perspective. *Journal of Early Adolescence.*

Lerner, R. M., Theokas, C., & Jelicic, H. (in press). Youth as active agents in their own positive development: A developmental systems perspective. In W. Greve, K. Rothermund, & D. Wentura (Eds.), *The adaptive self: Personal continuity and intentional self-development*. Göttingen, Germany: Hogrefe/Huber Publishers.

Lerner, R. M., & Walls, T. (1999). Revisiting individuals as producers of their development: From dynamic interactionism to developmental systems. In J. Brandtstädter & R. M. Lerner (Eds.), *Action and self-development: Theory and research through the lifespan*. Thousand Oaks, CA: Sage.

Leventhal, T., & Brooks-Gunn, J. (2004). Diversity in developmental trajectories across adolescence: Neighborhood influences. In R. M. Lerner & L. Steinberg (Eds.), *Handbook of adolescent psychology* (pp. 451–486). New York: Wiley.

Lewontin, R. C. (2000). *The triple helix*. Cambridge, MA: Harvard University Press.

Magnusson, D. (1999a). Holistic interactionism: A perspective for research on personality development. In L. A. Pervin & O. P. John (Eds.), *Handbook of personality: Theory and research* (2nd ed., pp. 219–247). New York: Guilford.

Magnusson, D. (1999b). On the individual: A person-oriented approach to developmental research. *European Psychologist, 4,* 205–218.

Magnusson, D., & Stattin, H. (1998). Person-context interaction theories. In W. Damon (Series Ed.) & R. M. Lerner (Vol. Ed.), *Handbook of child psychology: Vol. 1. Theoretical models of human development* (5th ed., pp. 685–759). New York: Wiley.

McLoyd, V. C. (1998). Children in poverty: Development, public policy, and practice. In W. Damon (Ed.) & I. E. Sigel & K. A. Renninger (Vol. Eds.), *Handbook of psychology: Vol. 4. Child psychology in practice* (5th ed., pp. 135–208). New York: Wiley.

Michel, G., & Moore, C. L. (1995). *Developmental psychobiology: An interdisciplinary science.* Cambridge, MA: MIT Press.

Müller-Hill, B. (1988). *Murderous science: Elimination by scientific selection of Jews, Gypsies, and others. Germany 1933–1945* (G. R. Fraser, Trans.). New York: Oxford University Press.

Overton, W. F. (1998). Developmental psychology: Philosophy, concepts, and methodology. In W. Damon (Series Ed.) & R. M. Lerner (Vol. Ed.), *Handbook of child psychology: Vol. 1. Theoretical models of human development* (5th ed., pp. 107–187). New York: Wiley.

Plomin, R. (2000). Behavioural genetics in the 21st century. *International Journal of Behavioral Development, 24,* 30–34.

Proctor, R. N. (1988). *Racial hygiene: Medicine under the Nazis.* Cambridge, MA: Harvard University.

Roth, J. L., & Brooks-Gunn, J. (2003a). What exactly is a youth development program? Answers from research and practice. *Applied Developmental Science, 7,* 94–111.

Roth, J. L., & Brooks-Gunn, J. (2003b). What is a youth development program? Identification and defining principles. In R. M. Lerner, F. Jacobs, & D. Wertlieb (Series Eds.) & F. Jacobs, D. Wertlieb, & R. M. Lerner (Vol. Eds.), *Handbook of applied developmental science: Promoting positive child, adolescent, and family development through research, policies, and programs: Vol. 2. Enhancing the life chances of youth and families: Public service systems and public policy perspectives* (pp. 197–223). Thousand Oaks, CA: Sage.

Rowe, D. C. (1994). *The limits of family influence: Genes, experience, and behavior.* New York: Guilford.

Rushton, J. P. (1999). *Race, evolution, and behavior* (Special abridged ed.). New Brunswick, NJ: Transaction Publishers.

Rushton, J. P. (2000). *Race, evolution, and behavior* (2nd Special abridged ed.). New Brunswick, NJ: Transaction Publishers.

Sampson, R., Raudenbush, S. W., & Earls, F. (1997). Neighborhoods and violent crime. A multilevel study of collective efficacy. *Science, 277,* 918–924.

Scales, P., Benson, P., Leffert, N., & Blyth, D. A. (2000). The contribution of developmental assets to the prediction of thriving among adolescents. *Applied Developmental Science, 4,* 27–46.

Schmitt, D. P. (2003). Universal sex differences in the desire for sexual variety: Tests from 52 nations, 6 continents, and 13 islands. *Journal of Personality and Social Psychology, 85,* 85–104.

Smith, J., Maas, I., Mayer, K. U., Helmchen, H., Steinhagen-Thiesen, E., & Baltes, P. B. (2002). Two-wave longitudinal findings from the Berlin aging study: Introduction to a collection of articles. *Journal of Geronotology: Psychological Sciences, 57B,* 471–473.

Spencer, M. B. (1990). Development of minority children: An introduction. *Child Development, 61,* 267–269.

Tobach, E. (1981). Evolutionary aspects of the activity of the organism and its development. In R. M. Lerner & N. A. Busch-Rossnagel (Eds.), *Individuals as producers of their development: A life-span perspective* (pp. 37–68). New York: Academic.

Tobach, E. (1994). . . . Personal is political is personal is political. . . . *Journal of Social Issues, 50,* 221–224.

Tobach, E., Gianutsos, J., Topoff, H. R., & Gross, C. G. (1974). *The four horses: Racism, sexism, militarism, and social Darwinism.* New York: Behavioral Publications.

Turkheimer, E., Haley, A., Waldron, M., D'Onofrio, B., & Gottesman, I. I. (2003). Socioeconomic status modifies heritability of IQ in young children. *Psychological Science, 14*, 623–628.

Vedantam, S. (2003, August 1). Desire and DNA: Is promiscuity innate? New study sharpens debate on men, sex, and gender roles. *The Washington Post*, p. A1.

Venter, J. C., Adams, M. D., Meyers, E. W., Li, P. W., Mural, R. J., Sutton, G. G., et al. (2001). The sequence of the human genome. *Science, 291*, 1304–1351.

Villarruel, F. A., Perkins, D. F., Borden, L. M., & Keith, J. G. (Eds.). (2003). *Community youth development: Programs, policies, and practices*. Thousand Oak, CA: Sage.

Weiss, R. (2003, September 2). Genes' sway over IQ may vary with class. *The Washington Post*, p. A1.

Will, G. (2003, September 21). Disconnected youth. *The Washington Post*, p. B7.

For Product Safety Concerns and Information please contact our EU representative GPSR@taylorandfrancis.com Taylor & Francis Verlag GmbH, Kaufingerstraße 24, 80331 München, Germany

Printed and bound by CPI Group (UK) Ltd, Croydon, CR0 4YY
01/05/2025
01858588-0001